Nikolai I. Lobachev:

The Foundations of Geometry

Works on Non-Euclidean Geometry

Translated by Svetla Petkova

With a Foreword by Ivan A. Karpenko

Edited by Vesselin Petkov

MINKOWSKI
Institute Press

Nikolai Ivanovich Lobachevsky
Born on 1 December 1792 in Makaryev, Russian Empire
Died on 24 February 1856 in Kazan, Russian Empire

© Minkowski Institute Press 2019

ISBN: 978-1-927763-24-7 (softcover)
ISBN: 978-1-927763-25-4 (ebook)

Minkowski Institute Press
Montreal, Quebec, Canada
http://minkowskiinstitute.org/mip/

For information on all Minkowski Institute Press publications visit our
website at http://minkowskiinstitute.org/mip/books/

Preface

This volume contains three works by Lobachevsky on the foundations of geometry and non-Euclidean geometry:

- Geometry (1823)

- Geometrical investigations on the theory of parallel lines (1840)

- Pangeometry (1855)

They are translated from their original Russian publication in the book: Н. И. Лобачевский, *Три сочинения по геометрии: Геометрия; Геометрические исследования по теории параллельных линий; Пангеометрия* (Гостехиздат, Москва 1956)
All figures are taken from the Russian publication.
The translation of these works turned out to be significantly more difficult than expected (which delayed the publication of the volume by over two years) due to a number of reasons. Here are two of them:

- There seemed to be problems in the original Russian text – for example, see the footnotes (Editor's Notes) on p. 25.

- Lobachevsky occasionally uses the same word for expressing different geometrical notions. For example, see the footnote (Editor's Note:) on p. 51: "By the word "величина" (translated here as"quantity" or "magnitude") Lobachevsky means "length", "area" or "volume". "

I would like to thank Svetla Petkova for the enormous effort put into translating and typesetting the book in LaTeX and Ivan A. Karpenko for writing such an informative Foreword.

2 October 2019

Vesselin Petkov
Minkowski Institute
Montreal

FOREWORD

The Revolution of Nikolai Lobachevsky

Nikolai Ivanovich Lobachevsky (1st of December (O.S. 20th of November), 1792 – 24th of February (O.S. 12th of February), 1856) – a remarkable example of a brilliant scientist, whose destiny had been full of drama due to the inability of most of his contemporaries –the scientific community of that time – to appreciate his contribution. This fact itself – the fact of the opposition to the genius and the inertia of the scientific community – is so important that here I will pay attention not only to an analysis of his key discovery in the field of non–Euclidean geometry but also to certain stages of the scientist's life path.

Lobachevsky came from a simple and poor family and the very fact that he eventually became the chancellor of the Kazan University and even the curator of the corresponding academic district at a certain point shows his great administrative talent. The beginning of Lobachevsky's path had been quite successful: the scheduled opening of the Kazan University where he had started learning and at the same time the invitation of brilliant teachers to the university who had most likely become a base for his scientific interest. He had been certainly lucky with the teachers – Martin Bartels, Franz Bronner, Kaspar Renner and others had been among his teachers. Of course, it is not just about the personal luck. It is well-known that the future scientist had made great efforts for his establishment as both administrator and scientist. He had succeeded more at the first if we talk about his formal recognition.

Despite his willfulness and disobedience, which have almost led to his dismissal and dispatch in the military, Lobachevsky had consistently climbed up the career ladder and reached perhaps the highest possible position in his environment. An achievement of a quick success – graduation with a master's degree, becoming an adjunct and then an extraordinary professor at the age of 24 and becoming the dean of the Faculty of Physics and Mathematics at the age of 28 – had been possible only because of his ability to establish good relations with high–level officials.

Indeed, Lobachevsky had had such an ability. In 1817 when the Ministry of Spiritual Affairs and National Education had been established (which had united science and religion), a serious crisis in education had started and raised the question of the closure of all universities (except the Moscow one), including the Kazan University. Foreign lecturers (Bartels and others) had been forced to leave the university due to the changes. Only by a miracle of the decision of Emperor Alexander I the university had not been closed but has become something like a spiritual seminary. Religion had come to the fore, it had become the main science, with which all the others must conform. In those disastrous conditions, Lobachevsky became the dean after reaching consensus with the officials in charge. It is significant that all that happened in the first quarter of the 19th century when Europe had been reaping the rewards of the Enlightenment and had become a progressive center of science, the place where rationality had finally prevailed over the mythological.

However, Lobachevsky's persistence in those conditions had allowed him to conduct research and aid the development of the university later on. Hardly after the period of obscurantism, he had been appointed a rector of the Kazan University – at the age of 35.

Unfortunately, that success had not accompanied his scientific and pedagogical work. Lobachevsky had published his course book with much difficulty – on the second attempt and in a modified version since reviewers had rejected the first one as a too loose interpretation of the Euclidean geometry.

In fact, Lobachevsky's research had suffered the same fate. Motherland had not acknowledged him as a scientist and, moreover, had mocked him (especially by the very influential mathematician of that time M.V. Ostrogradsky – Lobachevsky's "On the Foundations of Geometry" had received a negative evaluation from him).

His election as a corresponding member of the Goettingen Royal Scientific Society – due to the recommendation of K. F. Gauss – became the only fact of Lobachevsky's scientific recognition. Apparently being fully aware of the originality and importance of his "imaginary geometry" as he called it, Lobachevsky also published some of his works abroad. Thus, in 1837 Lobachevsky's article "Imaginary Geometry" appeared in an authoritative Berlin journal in French and in 1840 he published the pamphlet "Geometric Investigations on the Theory of Parallel Lines" in German. Gauss received the latter. He appreciated immediately Lobachevsky's idea since he himself had already been studying similar approaches to Euclidean geometry for a long time. Curiously, being an extremely acknowledged and influential mathematician of his time and considering non–Euclidean geometry an important and relevant step Gauss had not dared to publish the research on the topic and had not disclosed it to anyone except friends via letters (and had even warned against such activities). By his own admission, he had been afraid of criticism by the Boeotians (implying the representatives of that tribe to be considered stupid in the ancient world). His caution had strong ground because such authoritative scientists of his time (for example, in the case of Ostrogradsky and Lobachevsky) could be among the "Boeotians" and could seriously damage his reputation. Lobachevsky appeared to be more naive as he hoped that the scientific community abroad had not been as stagnant as in his motherland and he would have been understood there. Alas, his ideas could be understood only by a few people in the world at best.

Only Gauss's death and its result in the publication of his letters (where Lobachevsky had been mentioned too) have attracted the attention of the scientific community to the Russian mathematician and forced it to take his ideas seriously. It is significant that the very fact of publication of the Gauss's correspondence is not relevant to science, it bears purely publicist character, as a rather popular step for attracting the interest of the would–be scientific community. However, exactly that had worked. Unfortunately, Lobachevsky had also been no longer alive by that time. About ten years had separated him from his triumph.

The formation of Lobachevsky's theory is usually associated with his report "A Concise Exposition of the Principles of Geometry" as of 1826. In fact, that had been the birth day of non–Euclidean geometry. Here it is necessary to specify that Gauss and Janos Bolyai arrived at similar results (not identical but quite alike in variations of non–Euclidean geometry) in parallel to the works of Lobachevsky (mentioning parallels is rather ironic in this context). Moreover, the results of all the three could not have appeared out of the blue – the studies of many mathematicians and philosophers of different epochs (beginning with antiquity) and countries (Arab mathematicians have made considerable contribution) had prepared the ground. However, most likely, precisely Lobachevsky had become one of the first (if not the very first) who realized that non–Euclidean geometry can describe the physical reality.

It is necessary to give a brief overview of the problem background in order to comprehend Lobachevsky's contribution, i.e., the significance of his discovery. Let us recall Euclid's parallel axiom[1] from his "The Elements" , which had actually become a stumbling block:

That, if a straight line falling on two straight lines makes the interior angles on the same side less than two right angles, the two straight lines, if produced indefinitely, meet on that side on which are the angles less than the two right angles.

In modern literature its most common formulation is the one analogous to the fifth postulate authored by Proclus:

In a plane, given a line and a point not on it, at most one line parallel to the given line can be drawn through the point.

There are many other formulations of Euclid's parallel postulate. Replacing it with these formulations will lead us to the same Euclidean geometry and the fifth postulate is to be revealed as a theorem. Here are several curious versions:

At least one quadrilateral has all its angles right.

Straight lines parallel to the same straight line are also parallel to one another (authored by the mentioned above Lobachevsky's vocal critic Ostrogradsky).

Any triangle can be circumscribed by a circle (authored by Farkas Bolyai, the father of the mentioned above Janos Bolyai).

At least one circle has the ratio of its circumference to its diameter equal to π.

Many formulations had arisen because the parallel postulate differs from an axiom – a kind of basic knowledge on fundamental non-definable notions. It resembles more of a theorem. Moreover, the fifth postulate is not involved in the proof of the first 28 theorems of the "Elements" .

---

[1]Hereafter we shall consider both terms "postulate" and "axiom" equal.

Therefore, upwards antiquity many mathematicians had tried to prove it as a theorem deduced from other postulates. It seemed possible – in a similar way to the fourth postulate, which turned out to be a theorem. However, no one had succeeded (then and until now) in deducing it from the existing axioms (postulates). Therefore, many alternative axioms had been introduced for a derivation of the fifth postulate. However, these alternative axioms had been in fact the same fifth postulate and just as artificial.

Among all the predecessors of Lobachevsky, Gauss and Bolyai we should probably call Girolamo Saccheri the one who had come ever closer to the creation of non–Euclidean geometry. In 1733 he published a paper entitled "Euclid Freed of Every Flaw or a Geometric Attempt to Establish the Very Base of All the Geometry" . Saccheri planned the proof to the contrary. He decided to replace the fifth postulate with the opposite statement and to prove various theorems in the new theory. If the contradicting to the other theorems (or axioms) are to be proved in this geometry by means of the new axiom, this will mean that the assumption of the fifth postulate negation is false and thus the validity of the fifth postulate will be proved.

Saccheri used the idea of Lambert's quadilateral[2] as the proof. Roughly speaking, this is a quadrilateral with three right angles. Saccheri considered all the same three hypotheses about the 4th angle of Lambert's quadrilateral. There are three options. Firstly, if the quadrilateral is a rectangle, which means the fourth angle is a right one. The other two Saccheri's options considered an obtuse and an acute angles. The right angle case leads us to the Euclidean geometry. Saccheri displeased the case of an obtuse angle because all the lines intersect. He traced a contradiction here – allowing the opposite to lead him to the intersection of the lines, but exactly this follows from the fifth postulate under the certain conditions.

The situation with an acute angle had been more complicated. After this assumption Saccheri had practically begun building the geometry of Lobachevsky. He discovered that in the "false geometry" (as he called it) may be lines that converge, but do not intersect. This resented him because it contradicts the nature of the straight line. However, this had not stopped him, he continued the research, made a mathematical error and actually got a contradiction. He delightedly announced a long–awaited solution to the problem. Unfortunately, Saccheri's work had been noticed only in 1889, that is, when non–Euclidean geometry had already been created and accepted.

Lobachevsky's discovery consists of his insight that if the fifth postulate is replaced by its negation, then (contrary to Saccheri's opinion) a full geometry will appear. This means that the fifth postulate is indeed an axiom and not a theorem. However, this is not at all an obvious axiom – that could be replaced. It is significant that the geometry of Lobachevsky

---

[2]Which had been firstly considered by Ibn al-Haytham in the 11th century in order to prove the fifth postulate but got its name due to the studies of Johann Heinrich Lambert (1766), where he tried to prove the fifth postulate again.

is compatible with an absolute geometry[3] (the geometry of Euclid without the fifth postulate) – in both theories there are the same first 28 theorems, proved without the fifth postulate. The Lobachevsky's statement proposed instead of the fifth postulate is formulated as follows:

through a point, which is not on a given line, go at least two lines that lie within one plane with a given line and do not intersect it.

This means that the geometry of Lobachevsky describes the space with negative constant curvature (for example, in Riemannian geometry the constant curvature is positive, in the Euclidean geometry it has zero value).

The importance of Lobachevsky's geometry is enormous. On the world view level, he proposed a new paradigm. On the purely scientific level, his theory had been applied to various fields of physics and mathematics.

At the previously mentioned first level (worldwide transformation of views), Lobachevsky's geometry means a denial of the intuitive concepts based on everyday experience. For centuries, the Euclidean geometry had been considered undeniable and exclusively true. Mathematicians had been confident in its axioms as pure a priori knowledge, independent of any experience. They have been considered as a reflection of some higher knowledge, changeless, as eternal truths of pure reason. However, the approach of Lobachevsky (Gauss, Bolyai, Riemann and some others) had shown the opposite. The axiom of parallel lines is a vivid example of this – in fact, it had been a consequence of everyday experience, the result of everyday observations, just like other axioms. Parallel lines exist, and it seems obvious that only one line parallel to the given line can be drawn through a single point lying outside this line. However, Lobachevsky had questioned the second part and Riemann the first.

It is important to understand that Lobachevsky had not questioned the experience data. He meant the doubtfulness of the everyday observation data. The theory must be confirmed by practice, but this must be a scientific experiment that goes beyond the "visual", the "obvious" , etc. In other words, if anything seems to be indisputable to us by virtue of its intuitive acceptability, it is not true. Since intuition itself is based on everyday life practice. The scientific experiment allows you to go beyond the "common sense" – to see the world as it exists in reality. Both Lobachevsky and Gauss have seriously believed that their non–Euclidean geometries could describe the real physics of space, and Euclidean geometry would turn out to be an individual case.

As it has been already mentioned, the geometry of Lobachevsky had not only a symbolic value – as the paradigm shift, but had also been applied in science many times. The mathematician Eugenio Beltrami proposed several working models of Lobachevsky geometry – a pseudo-sphere, a projective model (independently described by Felix Klein) and a conformal disk model (or the Poincaré disk model named after the mathematician Henri Poincaré, who found the connection between this model and the problems of the theory of functions of a complex variable). We can also mention the

---

[3] In contrast to the spherical geometry and the geometry of Riemann.

connection between Lobachevsky's geometry and the kinematics of the special theory of relativity.

Lobachevsky geometry models construction, the proof of its consistency become possible. More precisely, the geometry of Lobachevsky is consistent if and only if the Euclidean geometry is consistent. And Euclidean geometry, as Hilbert had shown, is consistent if and only if formal arithmetic is consistent. Kurt Gödel in his famous theorem (1931) had shown that the latter problem cannot be solved by finite methods. But in 1936 Gerhard Gentzen proved the consistency of formal arithmetic using transfinite induction. Thus, the geometry of Lobachevsky is proven consistent by transitivity.

In addition, I would like to mention one more, and perhaps, one of the most important applications of Lobachevsky's geometry. The widely used today in various cosmological applications (and not only in them) anti–de Sitter space is a pseudo–Riemannian analogue of an n–dimensional hyperbolic space, that is, a space of a constant negative curvature, described by Lobachevsky's geometry. This connects Lobachevsky's geometry with the general theory of relativity – the anti–de Sitter space is the most symmetric solution of Einstein's equations in a vacuum with a negative cosmological constant.

Here one could argue that our space is not the anti–de Sitter space, i.e. the latter does not describe the real world and is just a mathematical model. However, it helps to solve the tasks, which turn out to be applicable to the actual physical reality. The historical example of Lobachevsky's discovery has exactly shown that sometimes it is necessary to strive to go beyond the directly observable in order to make a breakthrough in science.

Ivan A. Karpenko, PhD
National Research University Higher School of Economics
School of Philosophy
e-mail: gobzev@hse.ru

CONTENTS

GEOMETRY
1823

Introduction

Geometry is the part of pure Mathematics in which methods of measuring space are described.

A geometrical body possesses only the property *extension* compared to the other bodies of nature.

Extension is a property of bodies to expand and come in touch with one another.

All bodies in nature have three extensions, but we can imagine only one extension – in *lines*, two extensions – in *surfaces*, and at the end all three – in *bodies*. So lines, surfaces, and bodies are Geometrical quantities. For this reason Geometry can be divided into three parts: about measurements of lines (Longemetry), about measurements of surfaces (Planimetry) and about measurements of bodies (Stereometry).

The assembling of geometrical quantities is realized through their mutual touching. Their measurement consists of filling of what is measured several times of the taken measuring unit or its parts, and connecting them through touching. In geometry the beginning and the end of a line have a common name – a *point*.

Representing a line as composed of points, we can add points along its extension.

Lines are *touching* one another, if they have a common point and one line remains only one side of the other. Two lines are *intersecting* if they have a common point and one line goes from one side of the second line to its other side. Lines are *merging* or coinciding when they are indistinguishable or when they have a common part. In the same way, surfaces are touching in points, or lines, they intersect in lines and are merging entirely or in parts. Geometry is ending where ways of measurements of all geometrical quantities appear. It is followed by the application of Analysis to Geometry.

1 MEASURING LINES

Straight lines are those lines which merge as long as they have two common points. *A plane* is a surface, which contains a straight line. *A circle* or, rather, *a circular line* is such a line on a plane, whose points are equally distant from a point not lying on the line. These distances are called *radii*, and their common point, from which they originate, is called *the center of the circle*. Circles are different, if they do not have equal radii and coincide only when their radii are equal and their center is common. The parts of a circle are called *arcs*, and by the property of the circle, coincide when they have a common center. A circle, which is considered as a whole, but considered separately from its center and its radii, is called also *circumference*.

For measuring a straight line we can use as a measure also a straight line, because such a measure, taken several times, can cover completely the measured straight line. Similarly, for measuring arcs we can use as a measure a part of the same circle. Consider a line A, which is straight or arc of a circle, and B is its measure and is also a straight line in the first case and an arc of the some circle in the second case. Let B fits n times in A with a remainder $B' < B$; now let B' fits n' times in B with a remainder $B'' < B'$; the remainder B'' fits in B' n'' times with a remainder $B''' < B''$; the remainder B''' fits in B'' n''' times with a remainder $B^{iv} < B'''$ and so on. Finally, there should be no remainder, or, what is the same, it is so small that our senses cannot perceive and which in any case cannot be measured. Let this happen for a line Q when it lies p times in a line P. Such a measurement can be presented by the following equations:

$$A = nB + B'; \quad B = n'B' + B''; \quad B' = n''B'' + B'''$$

$$B'' = n'''B''' + B^{iv}, \quad \text{and so on} \quad P = pQ.$$

From here

$$\frac{A}{B} = n + \cfrac{1}{n' + \cfrac{1}{n'' + \cfrac{1}{n''' + \cfrac{1}{n^{iv} + \text{and so on}...+\frac{1}{p}}}}}.$$

As all can be presented as one fraction, the magnitude of line A will be the product of that fraction and B, and the fraction itself is the content of line A with respect to B.

In order to measure both straight and curved lines we use a straight line, called a *meter*, and its tenths, hundredths, thousandths and so on parts. It is required for measurements that the measure, or its parts, should fit on the measured line until it covers it completely. But as curved lines cannot coincide with straight lines their measurement, strictly speaking, is impossible. What is understood in Geometry under measurements of curved lines is to divide curved lines into extremely small parts (small straight lines) as small as to obtain a better approximation; after that the sum of all the parts (all straight lines) is equal to the length of the curved line. This way of measurement is taken from the very way we measure things in nature: the circumference of a wheel is measured by a chain, whose parts are straight lines, which are used instead of the parts of a curved line. A measurement is considered more correct if the parts of the chain are smaller. But the most precise accuracy would be reached when instead of a chain we use a flexible wire , i.e., a chain with extremely small parts. In fact, the measurement of entire lines is the subject of Calculus, where deferential calculation shows a method for every line to find such a quantity to which the length of a given curve approaches as closely as it is divided into smaller parts, and as a result the line would be split in the smallest decimal parts; and this quantity will actually be considered as the real length of the curved line.

2 ON ANGLES

For comparison of arcs in a circle their magnitudes are expressed in four-hundredths parts of a circle[1] called *degrees*. These are divided again in tenths, hundredths, thousandths etc. parts. The degrees are expressed by whole numbers with the sign °, for example 25°; the parts of the degree are expressed by a decimal fraction.

Two converging straight lines must intersect only in one point, otherwise they would not be straight.

An *angle* is called an arc between two converging straight lines, the expressed in degrees, taken from the point of their intersection. It is clear that the number of degrees of such an arc remains the same regardless of the magnitude of the radius.

The angles we discuss here are called *linear* in order to distinguish them from other angles formed not between lines. An angle of 100° degrees is called *right* and two straight lines which form such an angle will be *perpendicular* to each other or: each line will be perpendicular to the other. When two straight lines form an angle of 200° degrees they form one straight line.

We can draw a perpendicular from each point of a line, because it is possible to draw a semicircle around each point from one side and then imagine that that semicircle is divided into two; then a perpendicular will pass through the middle (the division) point. Here we see that the perpendicular from a point on a straight line on a surface can be only one. The angle between two straight lines when they do not go beyond the intersection point, could be taken from two sides but usually the smaller of the two is taken. The angle between two lines when they continue beyond the point of intersection, could be taken also differently. If two lines a and b (Fig. 1) converge to a common point and their continuations beyond the point of intersection will be called A and B, respectively, then the angle between a and b and the angle between A and B differ only by their location, but have equal magnitudes; because the sum of each of these angles and the angle between a and B make an angle of 200°. Also the angles between a and B and between b and A will be equal. Such angles

---

[1]EDITOR'S NOTE: Lobachevsky used the metric angular unit adopted in France, as part of the metric system, after the revolution there – in terms of this unit, a right angle is 100° and the whole angle in a circle is 400°.

are called *vertical*. Two angles between a and b and between B and a or the angles equal to them (between A and B and between b and A, respectively) are angles between the same straight lines; that is why we should always denote which of the two is the angle between the intersecting lines.

Two planes intersect into a straight line which is the necessary condition for the definition of a plane. The angle between two planes or a *plane angle* is the angle between the perpendiculars lying on the two planes taken to the line of the intersection of the planes. The extension of the planes beyond the intersection line will form vertical angles, which, like in the case of straight lines, are equal. Here we also should denote, which of the two angles will be the angle between the two intersecting planes. When the angle between two planes is right, then the planes will be perpendicular to each other.

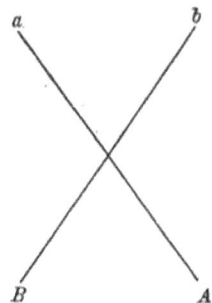

Fig. 1

Linear and plane angles are called *acute* when they are $< 100°$; and *obtuse* when they are $> 100°$.

The angle of a straight line with a plane is the angle between the line and the intersection line of the plane and a second plane, which contains the line and is perpendicular to the plane.

The surface of a *sphere* is a surface whose points are at the same distance from a point – the center of the sphere. The equal distances are the *radii* of the sphere; two radii on a straight line form the *diameter* of the sphere. Parts of the surface of the sphere coincide with the surface of the sphere everywhere when they have a common center. This property of the parts of the surface is similar to the property of arcs of a circle, because it is possible to measure the parts of the surfaces of spheres in the similar way.

Surfaces are bounded by lines. Planes can be bounded either by straight or curved lines. Surfaces bounded by straight lines are called *polygons*, whose straight lines are called *sides*. Less than three sides cannot bind a plane; if we have three sides, then the plane is called a *triangle*, if four, then – a *quadrangle* and so on depending on the number of sides. The parts of the polygons are: sides, angles, whose number is always equal to the number of sides, and *area*, whose name is accepted for denoting the whole surface of the polygon.

Triangles, with two equal sides are called *isosceles* and triangles with three equal sides - *equilateral*. In the isosceles triangle the unequal side is called the *base*, and the two equal sides are called *legs*. A triangle which has one right angle is called a *right* triangle, the side opposite to the right angle is called the *hypotenuse* and the sides adjacent to the right angle are called *legs* (or *catheti*). A plane bounded by a circular line is called a *circle*, a plane bounded by the chords and the arc – *segment*, a plane bounded by the arc and the two radii – *sector*.

Bodies are bounded either by planes or by curved surfaces. If a body is bounded only by planes then the parts of such a body will be: the very planes – *sides*, *plane* angles, *solid* angels and *volume* – the space inside the bodies. The sides of the bodies intersect one another into straight lines, which are called *edges*.

If a body is bounded by triangles, forming a solid angle, and other triangles and polygons, which complete the bounding of the body, then such a body is called *pyramid*. It will be *triangular pyramid*, if all of its sides are triangles; *polygonal pyramid*, if it has a polygon, which is denoted always by the name of: the *base* of the pyramid.

3 ON PERPENDICULARS

The method of drawing perpendiculars from points to lines or to planes and drawing a perpendicular from a point on a line or on a plane is taken from the properties of isosceles triangles, which are the following:

In an isosceles triangle the angles opposite to the equal sides are equal. Let in $\triangle ABC$ (Fig. 2) the side $b = c$, and by covering such a triangle with it we put point A on A, the line b on line c, and since the angle stays the same, then c will cover b; then by the equality of b and c, point C will coincide with point B, point B will coincide with point C, the side a will cover side a and therefore $\angle B$ will be covered by the angle C, e.g., $\angle B = \angle C$.

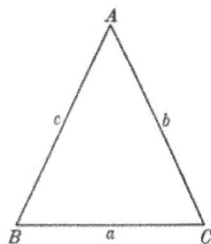

Fig. 2

On the contrary, if in a triangle two angles are equal, then the opposite sides are equal. Let in $\triangle ABC$ $\angle B = \angle C$, then covering $\triangle ABC$ with it, starting with putting point C on point B, the line a on a, and connecting the planes by the equality of the angles B and C, side b must coincide with c, side c must coincide with b, therefore they will intersect at point A, which shows, that $b = c$. Such triangles with two equal angles will be therefore isosceles.

Equilateral triangles, which are also isosceles triangles, should have all three angles equal, and vice versa: triangles, whose three angles are equal, should be equilateral.

In an equilateral triangle, a line passing through the apex and the middle of the base, will be perpendicular to the base and will divide the apex' angle into two equal angles. If a line passes through the apex and divides the angle into two equal angles then it will be perpendicular to the base and will divide it into two equal parts. If the line is perpendicular to the base, then it will divide both the base and the apex's angle into two equal parts. In $\triangle ABC$ (Fig. 3), if $b = c$, and therefore $\angle B = \angle C$, if D is the middle point of the base a, then $\triangle ADC$ will cover $\triangle ADB$, when

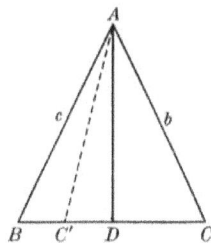

Fig. 3

11

the side DC is put on DB , the side b on c, so $\angle ADB = \angle ADC = 100°$, $\angle BAD = \angle DAC$.

When AD divides $\angle A$ into two equal parts, then again $\triangle ADC$, covers $\triangle ADB$, putting them one over the other by placing side b on side c, keeping AD as a common line, hence $BD = DC$, $\angle ADB = \angle ADC$.

If AD is a perpendicular to BC, then superimposing $\triangle ADC$ on $\triangle ADB$, keeping AD common,
the side DC will go on DB and will not finish in
C', otherwise it would be that $\angle AC'B = \angle B$, but
as $\angle ACD$ [together with $\angle AC'D$] is also equal
to $\angle B$, then AC' should be perpendicular to BC
together with AB, AC and AD, which is not possible; otherwise the extensions of AC and AB on
the other side of BC will converge again.

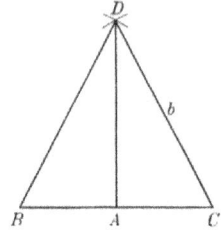

Fig. 4

And so, in order that we draw a perpendicular
from point A to a line (Fig. 4), we should take the
equal lines AB and AC, after that the line $b > AB$, regard points B and C
as centers, and draw circles around them, which will intersect somewhere
at point D, and AD will be perpendicular to BC.

In order that we draw a perpendicular from
point A to BC (Fig. 5), we should take somewhere [on the given line] a point B and draw a
circle around A with a radius AB. If the circle
does not go on the other side of the line BC we
should take a radius which is greater than AB
– then it will certainly happen. A circle, going
from one side to the other of the line BC, will
intersect the line in two points B and C. From
the points B and C, we draw equal lines BE and
EC, after which we draw the line AE we form
two triangles AEB and AEC, which are equal,
because they cover each other, when we put the
line EC over EB and AC over AB, and therefore $\angle BAD = \angle CAD$ and AD is perpendicular
to BC.

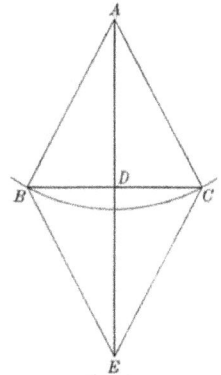

Fig. 5

A perpendicular to a plane is called such a line, which is a perpendicular to all lines, drawn in the plane through the point of intersection.
Obviously, if a line is perpendicular to two lines in the plane, then it will
be perpendicular to all lines, and therefore it will be perpendicular to the
plane.

Let (Fig. 6) AB be perpendicular to CC' and DD' at the point of
their intersection B, then AB will be perpendicular to any third EE'. We
remove the lines AB, BC, BE, BD, keeping their mutual configuration,
then we transport them to other lines in such a way that point B stays on
its place. After that we put the plane ABD on the plane ABD'. By the
equality of vertical plane angles, the plane ABC will cover the plane ABC',

by the equality of right angles the line BD lies on BD', BC on BC' and therefore the plane DBC lies on $D'BC'$. At the end, by the equality of the angles EBD and $E'BD'$ the line BE coincides with the line BE', that is, the angles ABE and ABE' will be equal, then each of them is a right angle.

Two planes intersect in a line, which is perpendicular to a third plane, to which both of them are perpendicular. When the planes AKB and CID are perpendicular to plane EF (Fig. 7), then the line of intersection GH of the first two planes will be perpendicular to the last (EF). We transfer the planes CD and AH, keeping their mutual configuration, on the planes CD and GB. We leave the point H at the same place, line HD is put on

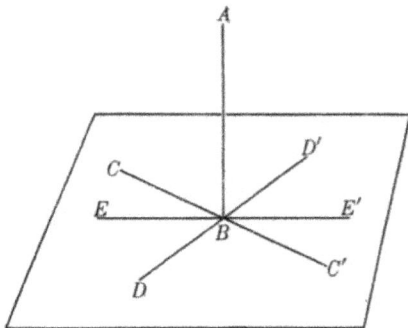

Fig. 6

HI, the plane EF again on EF, then the equality of the angles KHD and BHI will force the line HK to coincide with the line HB. At the end, the equality of the right angles requires that the plane GD coincides with the plane CH, the plane KG coincides with the plane GB, and therefore their common line of intersection will be GH, the angles GHI, GHD, GHK, and GHB are equal among themselves and all are right angles.

A perpendicular to the line of intersection of two planes perpendicular to each other is perpendicular to one plane, when it lies in the other plane. Otherwise, a third perpendicular plane to the first plane in the point of intersection with the second, passing through the foot of the perpendicular, will produce another perpendicular to the line of intersection of the first two planes.

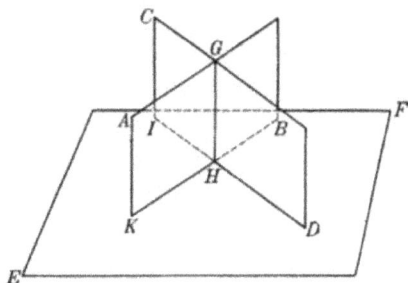

Fig. 7

A plane, passing through a perpendicular to another plane, is also perpendicular to this plane. Because the line, drawn in the second plane using the foot of the perpendicular under right angle to the line of the intersection of the two planes, makes a right angle with the perpendicular, which will be the angle of the two planes.

From what was proved above about perpendicular planes and perpendicular lines, a rule follows, which shows how to draw a perpendicular from a point on a plane and how to draw a perpendicular to a plane from a point.

In order that we put a perpendicular on a plane we must
take any line in the given plane
not lying on a given point on the
plane. Then we draw a perpendicular to this line from the given
point, and another perpendicular
not lying on the plane. We draw
a plane through the last two lines
and in the plane we draw a perpendicular line to the line on the given
plane, which line has been drawn

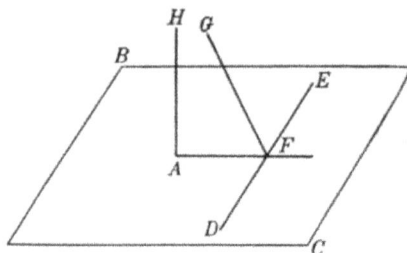

Fig. 8

from the given point. Consider a point A on the plane BC (Fig. 8), we
draw in the plane an arbitrary line DE, after that we draw a perpendicular AF from point A to it and another perpendicular FG not lying in the
plane BC, then the plane drawn through AF and GF will be perpendicular to the line DE, therefore also perpendicular to the plane BC, because
the perpendicular AH to AF in the plane AFG will be perpendicular to
the plane BC.

In order to draw a perpendicular from a point on a plane we
should draw arbitrarily a line in the
plane, we then draw to it a perpendicular from the given point, another perpendicular is drawn in the
plane through the foot of the first
and to this last line draw a perpendicular from the given point. Consider a point A and a plane GH
(Fig. 9). Let us draw an arbitrary
line BC in GH, then from A we
draw to it the perpendicular AD,

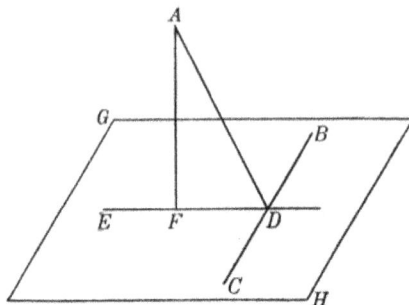

Fig. 9

through the point D we draw another perpendicular ED in the plane GH
and to this last line we draw the perpendicular AF, which will be a perpendicular to plane GH by the same reason given above.

From here we find a method to build perpendicular planes. If we have
a line on a plane and we need to draw through it a perpendicular plane to
the initial plane, then we need to draw a perpendicular line lying on the
plane to the given line, and draw a perpendicular out of the plane at the
point of intersection of the first two lines; through the last line and the
initial line we draw a plane, which is the one we needed.

If we have to draw a perpendicular plane to another plane through
a line out of the plane, then we should only draw perpendiculars from
two points on the given line and through them draw the plane, or, when
the given line intersect the plane, at the point of intersection we draw
a perpendicular line on the plane and another perpendicular to the last

on the plane, through which the given plane passes. The last method is shorter but it is supposed that the line intersects the plane, or can intersect it if it is sufficiently extended. The proof is self-evident.

Perpendiculars are the shortest distances between a point and a line and between a point and a plane. If AB is perpendicular to CD (Fig. 10), then AB is
smaller than any other line AE, drawn
from A to CD. On the other side of AB we
imagine that a triangle $\triangle ABF$ is equal to
$\triangle ABE$, for which we need only to transfer
BE to the other side of the point B. The
circle, drawn from point A with a radius
AE, will pass through E and F and will

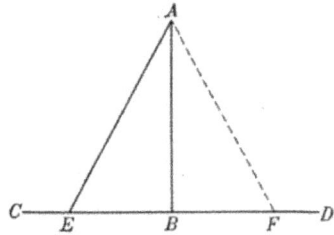

Fig. 10

not intersect the line CD in a the third point, otherwise we would have an isosceles triangle, in the middle of which we could draw another perpendicular from point A to the line CD. It is seen from here, that AB should be extended in order that it goes to the circle, drawn by the radius AE and therefore $AB < AE$.

A perpendicular from a point to a plane will be the shortest distance as well, otherwise it will turn out that the perpendicular is longer than an inclined line.

4 MEASUREMENT OF SOLID ANGLES. ABOUT REGULAR POLYGONS AND BODIES

The surface of a sphere, like a circle, is divided into 400 equal parts by planes, going through a radius. These parts are called *degrees* and are divided in tenths, hundredths and so on.

Part of the surface of a sphere, formed by planes, passing through the center, and whose magnitude is measured in degrees, is called a *solid angle*. Planes, forming a solid angle, will be called *sides* of the angle and depending on the number of these planes it is called *three-sided, four-sided* and so on.

Like polygons could be formed by adding and subtracting triangles, multi-sided solid angles can be formed from three-sided angles. Because measuring solid angles in general boils down to measuring only three-sided angles.

The intersection of a plane and the surface of a sphere results in a circle.

The perpendicular CE (Fig. 11) from the center of the sphere C to the plane ADB, which intersect the surface of the sphere in the lines ADB, could not fall out of the sphere, for example at point F, because otherwise the line AF drawn on the plane inside the sphere, would have two points of intersection A and B on the surface of the sphere, which together with the center C would form an isosceles triangle, and in it, one could draw a perpendicular to the line AF in addition to CF. Two points A and D, taken arbitrarily on the line ADB, form together

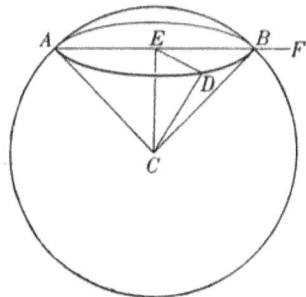

Fig. 11

with the center two triangles ACE and DCE, which will be placed next to each other on one plane, keeping the line CE common, will form an isosceles triangle, where the line CE from the apex will be perpendicular to the the base, and therefore $AE = ED$. As all similar lines AE and ED are equal, then the line ADB is a circle, and the point E is its center.

All solid angles bounded on the surface of the sphere by circles whose

17

radii are equal with the radius of sphere and therefore those circles have the greatest radii. For this reason the circles whose centers coincides with the center of the sphere are called the greatest circles.

The *apex* of a solid angle is the common point of intersection of the sides. The extension of the sides beyond the point of intersection form another solid angle, which, together with the first angle, are called *apex solid angles*. *Isosceles* and *equilateral* [three-sided] solid angles are those which have two or all three sides equal.

The apex solid angles are equal. It is necessary to show this only for three-sides angles, because the planes, dividing such an angle into three-sided angles, divide the extention beyond the apex – another angle – into the same number of three-sided angles.

If in the three-sided solid angle $CABD$ (Fig. 12) the sides ACB and ACD are equal, then their apex angle $CA'B'D'$ is equal; because putting $\angle ACB$ over $\angle A'CD'$, the equality of the plane angles between CAB and

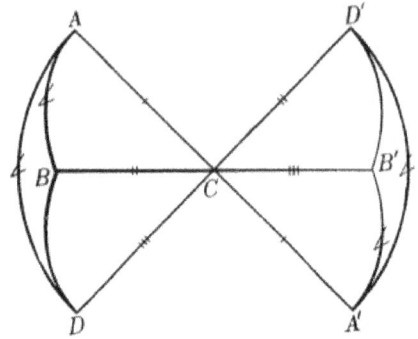

Fig. 12

CAD will make the plane ACD to cover the plane $A'CB'$ and also $\angle ACD = \angle A'CB'$, $\angle ACB = \angle A'CD'$, then the line CB will coincide with CD', CD with CB', so one angle will completely fill the other.

If the solid angle $CABD$ (Fig.13) is not isosceles, then we draw a plane through the points A, B, D, and also draw to it a perpendicular CE, then the triangles AEC, BEC, DEC will be formed with equal angles ACE, BCE, DCE; because they, having a common line CE, and being placed one next to another in one plane, form

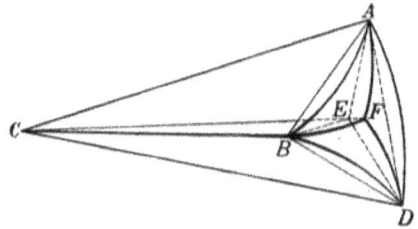

Fig. 13

an isosceles triangle, in which the line through the apex is perpendicular to the base. It follows from here that CE, when extended to F on the surface of the sphere, gives equal arcs AF, BF and DF, i.e., the solid angle ABD is divided into three isosceles solid angles AFB, BFD, DFA, which should be added or they should be subtracted from others, depending on whether F is inside or outside the solid angle. Now, when the solid angle is divided into three isosceles triangles with the help of three planes, the corresponding to it apex angle will be divided by the same planes into equal angles, and therefore the apex solid angles are equal.

Imagine a part of the surface of a sphere, cut off by the planes of the three-sided solid angle, whose apex is at the center of the sphere. The lines a, b, c (Fig. 14) are lines of intersection of the planes with the surface of the sphere, points A, B, C will be the points of intersection of the lines between themselves, taken in such a way that A will be against a, B against b, C against c. We extend a in both direction on the sphere's surface, until we complete a whole circle. We extend the arcs b and c beyond the point of their intersections A, until they reach the circle of a. We obtain four solid angles X, Y, Z, and

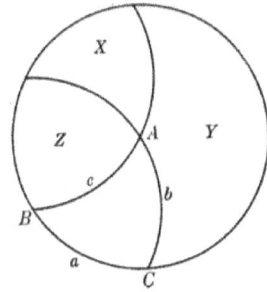

Fig. 14

$\bigwedge ABC$. Denote by $\mathbf{4}A$ the plane angle in $\bigwedge ABC$ which is opposite the arc a, and denote by $\mathbf{4}B$, $\mathbf{4}C$ two other angles opposite the arcs b and c. It is easy to see, that

$$\bigwedge ABC + Y = \mathbf{4}B;$$

$$\bigwedge ABC + Z = \mathbf{4}C.$$

The same applies to the apex solid angle with the angle X, and as a result instead of it the angle X itself alone gives

$$\bigwedge ABC + X = \mathbf{4}A.$$

At the end the sum of four solid angles X, Y, Z, and $\bigwedge ABC$ is 200°, which added to the previous three equations gives:

$$\bigwedge ABC = \frac{\mathbf{4}A + \mathbf{4}B + \mathbf{4}C - 200°}{2}.$$

In this way, we define the solid three-sided angle with the help of plane angles.

From here it is not difficult to conclude how to define any solid angle from the plane angles which are forming it. If we divide the solid angle into three-sided, following the above method, we see, that for a number of sides n the magnitude of the solid angle will be:

$$\frac{1}{2} \times \{\text{sum of plane angles} - (n - 2) \, 200°\}.$$

From here we draw important conclusions about regular bodies. But in order to fully cover the subject first we will consider *regular polygons*. We call regular polygons those, whose sides and angles are equal. We can form a regular polygon with n sides, when we divide a circle into n equal parts and connect the points of intersection by lines. Then the center of the circle is called also *center* of the regular polygon. On the contrary, in any regular polygon

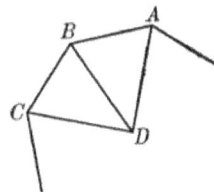

Fig. 15

there is a center. Divide one of the angles by a line [into two equal parts]. Such a line will pass through the whole polygon and will divide it into two equal parts. From here it follows, that another similar line will intersect the first one inside the polygon. Let AD divide the angle A into two equal parts (Fig.15), BD divides the angle B, which is equal to A, into two equal part, and D is their point of intersection. In $\triangle ABD$ the two sides AD and BD will be equal, the third line CD forms a $\triangle CBD$, equal to $\triangle ABD$, as long as the angle $C = B = A$ and the line $BC = BA$. Because, putting $\triangle ABD$ on $\triangle DBC$, leaving BD a common side, the line BA must fail on BC with point A on C, therefore $AD = DC$ and $\angle DCB = \angle DAB = \frac{1}{2} \angle A$. So all distances from D to the apexes of the angles in a regular polygon will be equal to AD; then D is the center of the polygon.

So, it is possible to have regular polygons with any number of sides and there is a center in every regular polygon.

A *regular body* is called a body which is bounded by regular polygons and whose solid and plane angles are equal. It is needed also that all polygons of the body are equal, and that equal number of polygons fit in all solid angles of the body.

Solid angles whose plane angles are equal and have equal sides can be called *regular solid angles*. In them the planes, dividing the plane angles ino two equal parts, intersect in common lines – the *axes* of the angle, and each of them divides the solid angle into two equal parts, and therefore, coming out of the solid angle, it either coincides with the edge or divides the side into two equal parts. The axes are inclined to the edges under equal angles.

In the regular body the plane, dividing one of the plane angles into two equal parts, gives when intersects the surface of the body either a regular polygon, or such a polygon where the sides going through one side of the body and the angles through one angle of the body are equal. In the first case it is not difficult to see that the center of the polygon will be a *center* of a regular body, that is, such a point, which is situated at equal distance from the apexes of the solid angles. In the second case, connecting the ends of every two adjacent sides by a line we again get a regular polygon, whose center will be the center of a regular body.

Let the number of the sides of a regular body is n, every side is a regular polygon with m sides, in every solid angle fit t polygons.

Drawing a sphere from the center of a regular body, then each side will correspond at the center a regular solid angle with m sides, in which every plane angle $= \frac{400°}{t}$, and so the whole solid angle $= \frac{1}{2}\{m \times \frac{400°}{t} - (m-2) \times 200°\}$. As such an angle should be together with nth part of $400°$, then

$$n = \frac{4t}{2m - (m-2) \times t}.$$

Numbers n, t, m should be positive integers and since $m \geq 3$ than also $t \geq 3$. Let $m = 3 + p$, where therefore $p = 0$, or $p > 0$, then

$$n = \frac{4t}{6 - t - (t-2)p}.$$

From here it is obvious, that t should be less than 6. If we take $m = 6 + q$, then n is

$$\frac{4t}{12 - 4t - q(t - 2)},$$

and since $12 - 4t$ could be only 0 or a negative number, and $t - 2$ is always a positive number, then q must be negative, that is, m is always smaller than 6.

And so all suggested numbers m and t are described in the table:

m=3 t=3 n=4   body is called *tetrahedron* (four-sided)
m=3 t=4 n=8   ”   ”   *octahedron* (eight-sided)
m=3 t=5 n=20 ”   ”   *icosahedron* (twenty-sided)
m=4 t=3 n=6   ”   ”   *cube* (six-sided)
m=4 t=4 denominator in n becomes negative, so such a body is impossible
m=5 t=3 n=12 body is called *dodecahedron* (twelfe-sided)
m=5 t=4 denominator in n is negative and therefore such a body is impossible
m=5 t=5 n is also negative and such a body is impossible

Regular polygons, as we saw, could be with any number of sides. On the contrary, regular bodies can be only five.

What is also different in regular polygons, compared to regular bodies, is that the number of their sides is equal to the number of their angles, whereas in regular bodies this is not so. Let r denote the number of angles in a regular body, keeping the previous notations, $n \times m \frac{400°}{t}$ will represent the sum of all plane angles, which will be formed, when from the center of the regular body we draw planes through all sides of the polygon. On the other hand, this sum should give $r \times 400°$, therefore $r = \frac{nm}{t}$. So

angles in tetrahedron...................... 4
angles in cube 8
angles in octahedron 6
angles in dodecahedron 20
angles in icosahedron 12

5 ON THE EQUALITY OF TRIANGLES

Geometrical quantities are called *congruent* when it makes no difference which one we will take. That they are congruent can be verified by placing one over the other, then the lines and the surfaces of one coincide with those of the other. If two geometrical quantities are congruent only with respect to their parts, then they are *equal*. Sometimes the equality of only some parts of two quantities already ensures the congruence of the two; but it is necessary that those parts be corresponding to one another. Talking about the congruence of triangles we will not mention correspondence as a necessary condition, which should be self-evident. Also, for brevity, instead of saying: the side, the angle of one triangle is equal to the side and the angle of another triangle we will say: the side is equal and the angle is equal.

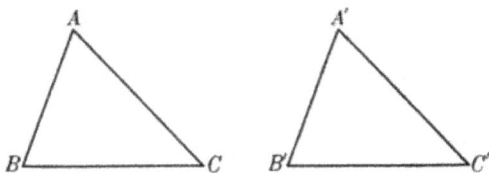

Fig. 16

Two triangles are congruent when they have a side and two adjacent angles equal. If in $\triangle ABC$ and $\triangle A'B'C'$ (Fig. 16) $BC = B'C'$, angle $B = \angle B'$, $\angle C = \angle C'$, then, putting the side BC over $B'C'$, by the equality of the angles, AB should coincide with $A'B'$ and AC with $A'C'$ and so they will converge at one point.

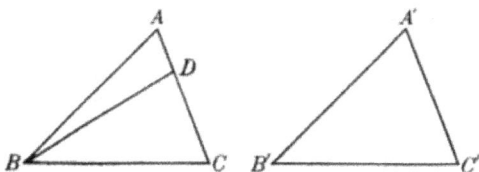

Fig. 17

Two triangles are congruent, when they have an equal side and two equal angles, one of which is opposite to the side. If in the triangle ABC and $\triangle A'B'C'$ (Fig. 17) the side $BC = B'C'$, the angle $C = \angle C'$, $\angle A = \angle A'$, in order that they are not congruent, it is necessary, that we can have $\triangle ABD$, when one is placed on the other on their equal side, where the sum of the two angles BAD and BDA is equal to 200°.

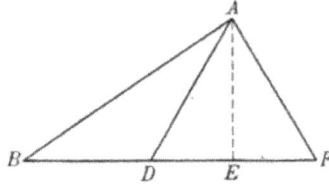

Fig. 18

But if in $\triangle ABD$ (Fig. 18) the sum of the angles B and D is equal to 200°, then neither B, nor D could be right, otherwise they would both be right angles, and the sides AB and AD would be two perpendicular [to BD]. If, B, for example, is an acute angle, then D is therefore obtuse, then the perpendicular AE to BD should fall outside of $\triangle ABD$ on the side of BD. Assuming $DE = EF$, introducing AF, then we get two equal triangles AED and AEF, therefore $\angle ADE = \angle AFD$, and since $\angle ADF =$ angle B, then $AB = AF$ and the point E should be in the middle of BF, which cannot be.

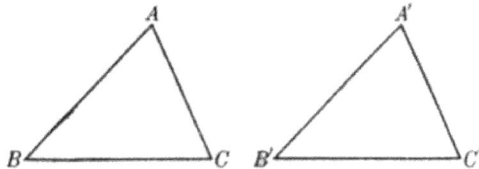

Fig. 19

Here it is mentioned in the proof that the perpendicular falls from one side of the angle to the other, so it is facing the acute angle. The reason for that is that when two perpendiculars could not converge, then it is less probable that a line cannot converge to the perpendicular, while facing the obtuse angle.

Two triangles are congruent, when they have two sides and the angle between them equal. If in $\triangle ABC$ and $\triangle A'B'C'$ the side $AB = A'B'$, $AC + A'C'$, the angle $A = \angle A'$ (Fig. 19), then putting one over the other one of their equal sides, for example AB on $A'B'$, the other side AC should lie on $A'C'$ by the equality of the angles A and A'. In such a way one triangle will cover the other.

Two triangles are congruent, when they have two equal sides and an equal angle that is opposite to the longer of the two equal sides. In $\triangle ABC$ and $\triangle A'B'C'$ (Fig. 20) let $AB = A'B'$, $AC = A'C'$, $\angle B = \angle B'$ where $AC > AB$. We superimpose $\triangle A'B'C'$ on $\triangle ABC$ the side $A'B'$ on AB,

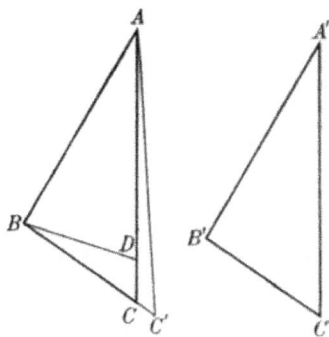

Fig. 20

then $B'C'$ will be superimposed on BC, but let it end in C' not in C. We will get a $\triangle ACC'$ which is isosceles, where the perpendicular to the base CC' will pass through its middle and therefore the angles at the base will be acute, the angle ACB, that is, the angle C of the triangle ABC will be obtuse.[1] Let $AD = BA$ and as $AC > AB$, then point D should be between A and C; therefore $\triangle ABC$ will be divided into $\triangle ABD$ which is isosceles and another $\triangle BDC$. In $\triangle ABD$ the angle ADB is acute, and in $\triangle BDC$ the angles BDC and BCD are obtuse.[2]

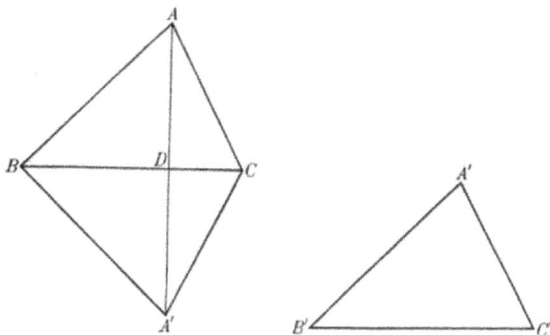

Fig. 21

Now, if we imagine perpendiculars in the points D and C, then they should go inside of $\triangle BDC$, and so they will intersect.

If point C' of the triangle $A'B'C'$ is inside of $\triangle ABC$, then, on the contrary, superimposing triangle ABC on $\triangle A'B'C'$, the side AB lies on $A'B'$, then point C will be outside of $\triangle A'B'C'$, which should not happen. Therefore point C can only go in C'.

Triangles are congruent when they have three equal sides. Let $AB =$

---

[1]EDITOR'S NOTE: It seems there is a problem in the Russian text "...угол же ACB, то есть угол C треугольника ABC будет тупой;" obviously, the angle C is not obtuse.

[2]EDITOR'S NOTE: Again, there seems to be a problem in the Russian text "В $\triangle ABD$ угол ADB острый, а в $\triangle BDC$ углы BDC и BCD тупые;" obviously, the angle BCD is not obtuse.

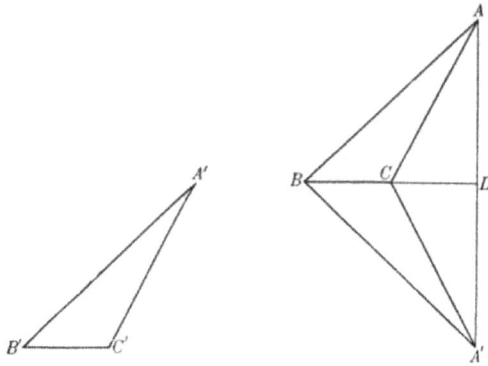

Fig. 22

$A'B'$, $AC = A'C'$, $BC = B'C'$ (Fig. 21) in the triangle ABC and $\triangle A'B'C'$. Arranging $\triangle A'B'C'$ to ABC, in order that the side BC is a common side of both and that equal sides intersect in the points B and C. Connect A and A' with a straight line. Then two isosceles triangles will be formed – ABA' and ACA' – in which $\angle BAA' = \angle BA'A$, $\angle CAA' = \angle CA'A$. If the point of intersection D of the lines AA' and BC will be between B and C, then $\angle BAA' + \angle CAA' = \angle A$; $\angle BA'A + \angle CA'A = \angle A'$, and therefore $\angle A = \angle A'$. If point D is outside $\triangle ABC$ and $\triangle A'B'C'$ (Fig. 22), then the difference of the angles BAA' and CAA' gives $\angle A$, the difference of $\angle BA'A$ and $\angle CA'A$ gives angle A', from where it follows again that angle $A = \angle A'$. When the angles A and A' in $\triangle ABC$ and $\triangle A'B'C'$ are equal, then the triangles must be congruent, because they have two equal sides and the angle between those sides is equal.

From the above cases of congruence of triangles it is clear that the parts of a triangle should be related to one another, and that there should exist a method to determine some parts of the triangle with the help of the others. Even here we can see that three parts of a triangle, with at least one side, determine the remaining three. This remark is important because, the determination of the unknown parts of a triangle is the subject of Trigonometry, and here therefore we can say in advance that the tasks of Trigonometry are to find the magnitudes of the three parts of a triangle when the other three parts are given. After that we will see that triangles are not always congruent, when they have only equal angles, therefore Trigonometry cannot give a method for determining the sides of a triangle, when only its angles are known. Since it is impossible to define lines with the help of angles, some Mathematicians wanted to regard that fact as the foundations of geometry. But such an idea is not sufficient, because dissimilar quantities could be dependent on one another.

6 ON MEASURING RECTANGLES

The measurement of planes is based on the fact that two lines converge, when they are situated on one side of a third line and when one is a perpendicular and the other is inclined under an acute angle facing the perpendicular. The lines AB and CD (Fig. 23) must converge if they are extended sufficiently, if one of them AB is a perpendicular to BC and the other CD – inclined relative to BC under an acute

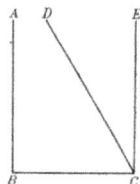

Fig. 23

angle C, which faces the perpendicular AB. But up to now we do not have a rigorous proof of that. Offered proofs can be at best called explanations, but could not be considered in the sense of Mathematical proofs. The best is the following:

Draw a perpendicular CE, let $\angle DCE = \frac{400°}{n}$, where n is a whole number. Place n such angles as DCE one after the other with their apexes together: we get a circle whose plane can be extended everywhere in all directions with no limits. When we get a line n times greater than BC and at the end of this line we draw perpendiculars, then the plane between the perpendiculars can extend only in one direction. From here it is seen that the plane between the lines DC and CE will increase much more with the extension of the lines DC and CE than the plane between AB and CE with the extension of the lines AB and CE and that at the end the first plane must be greater than the second. Then it will not fit any more between AB and CE, what else can happen as when CD intersects AB. When the angle DCE is a fraction of $400°$, then we make DC come closer to CE and an angle of $\frac{200°}{n}$ is formed, where n is a whole number. The line DC in its new position should converge to AB moreover, therefore, any other line CD will deviate further from CE.

From here it follows in the first place, that when a line is perpendicular to another, which is perpendicular to a third and the third is perpendicular to a fourth and when all of the four lines lie in the same plane, and the right angles are situated on one side of the lines, then the first line will be perpendicular to the fourth; so a quadrangle with right angles will be formed, which is called *rectangle*.

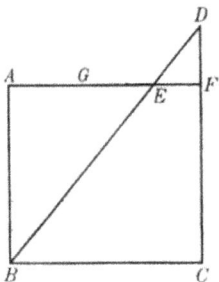

Let GA be a perpendicular to AB, AB to BC, BC to CD (Fig. 24), then any line BD drawn in the angle B should intersect both AG and CD. Suppose that BD first intersects AG at point E, then CD at point D, then the extension of AE will go in $\triangle BDC$ and cannot intersect either BC or BD but will intersect CD at some point F. We will have the same situation when BD first intersect CD, and then AG. The other assumption, that is, that BD converges together with AG and CD, would mean by itself, that CD and AG intersect. The angle at point F cannot be acute inside the quadrangle: otherwise the sides CF and BA would converge on the side BC, it cannot be obtuse either: otherwise the angle DFA would be acute and AB with CD would converge on the other side, which cannot happened, because AB and CD are perpendicular to BC.

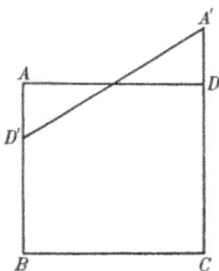

Fig. 24

In rectangles opposite sides are equal. If we suppose that in the rectangle $ABCD$ (Fig. 25) that the side AB is longer than CD, then covering the quadrangle $ABCD$ in such a way that CD goes over AB, and AB over CD, then point D should fall between the points A and B, somewhere at D'; on the contrary, point A would be outside of the quadrangle $ABCD$, somewhere at A', hence AD and $A'D'$ will intersect and will give two perpendiculars to a line from one point.

Fig. 25

Two rectangles are congruent, when their sides are equal. The easiest way to prove this is to superimpose the two rectangles.

When each side of the rectangle is divided into two equal parts, at the division point perpendiculars to the side are drawn, which will pass through the whole rectangle. Then the rectangle will be divided into congruent rectangles. It is clear that the perpendiculars should cut out rectangles; it is obvious as well that the first and the second rectangle have two equal sides, also the second with the third, the third with the fourth and so on until the last.

It follows from the above explanation, that when one of two adjacent sides in a rectangle is divided into n equal parts, the other into m equal parts and from the intersection points perpendiculars to the sides are drawn, then the rectangle will be divided into congruent rectangles, whose number will be the product of the numbers n and m.

Let in two rectangles A and B the adjacent sides CD, DE (Fig. 26) of the first relate to the adjacent sides FG and GH of the other, like the numbers p to n, q to m, then dividing CD into p equal parts: such parts in FG will be n. We divide DE into q equal parts, then such parts in GH will be m. The rectangle A will be divided in pq such rectangles, in B

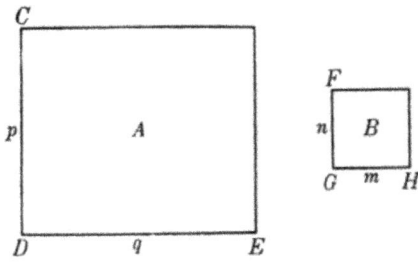

Fig. 26

their number will be nm; hence the content of rectangle A to B is equal to the content $\frac{pq}{nm}$.

A rectangle with equal sides is called a *square*. To measure flat surfaces we take a square as a unit, whose side is a linear unit. So the area of the rectangle is a product of two perpendicular sides, which are also called *hight* and *base*.

7 ON MEASURING TRIANGLES AND OTHER FIGURES

Take a rectangle, whose a and b are two perpendicular sides (Fig. 27), then draw a line inside the rectangle through the apexes of two opposite angles. Such a line is called *diagonal* in a rectangle as well as in every quadrangle; in this way we will have two right triangles, whose sides are a and b, and the hypotenuse is the diagonal and which triangles will be congruent because they have three equal sides. Hence each triangle is half

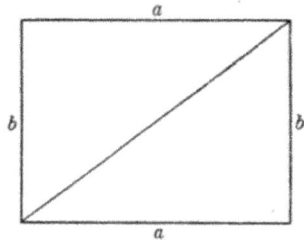

Fig. 27

of a rectangle. In the previous chapter we saw that the area of the rectangle is equal to the product of two perpendicular sides. Therefore the areas of the right triangles, which are produced by the diagonal in the rectangle, each is $=\frac{1}{2}ab$, and as all right triangles with sides a and b are congruent, then it follows from here, that the area of each right triangle is equal to a half of the product of the two sides.

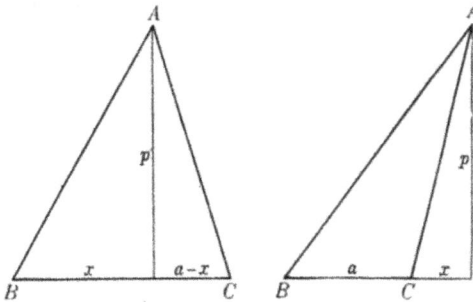

Fig. 28

Assume now $\triangle ABC$ (Fig. 28) is not a right triangle. From the apex of some angle, for example A, we draw a perpendicular to the opposite side a. As we assumed that the angle C is not right, then the perpendicular p cannot fall at point C. So we distinguish here two cases: either p is inside the triangle ABC or outside. In the first case p divides a into two parts:

31

x and $a - x$ and $\triangle ABC$ will be divided into two right triangles, so p and x will be sides of one of them, and p and $a - x$ sides of the other triangle. The area of the first triangle is $= \frac{1}{2}px$, the area of the second triangle $= \frac{1}{2}p(a - x)$, hence both of them will give the area of the $\triangle ABC = \frac{1}{2}pa$. In the second case we also have two right triangles: the first with sides p and x, calling x the nearest distance p to the angle C, the sides of the second triangle are p and $a + x$, that is, these are the perpendicular and the longest distance to p from the angle B. The area of the triangle ABC will be equal to the difference of the area of the greater right triangle and the smaller. As the area of the first $= \frac{1}{2}p(x + a)$, and the area of the second $= \frac{1}{2}px$, the area of $\triangle ABC = \frac{1}{2}pa$ is the same as it was in the first case.

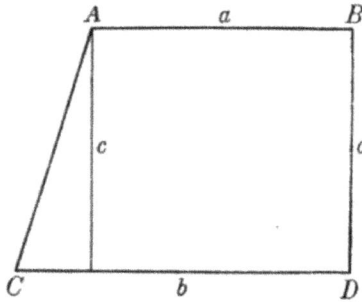

Fig. 29

The perpendicular p is called the *hight* of the triangle, and the side a to which p is perpendicular – the *base*. Extrapolating these names to right triangles as well, in a similar way one of the sides will be called hight of the triangle, the second – a base, and then the area of each triangle – both right and non-right, will be equal to a half of the product of the base and the hight.

Polygons can be always divided into triangles, and in such a way their areas can be measured. To do that we should know the hight and the base of all triangles, which are formed as a result of the division. If a polygon is given by its sides and angles, expressed by numbers, then it is better to find by calculation the hight and the base of the triangles, which comprise it, which needs the help of Algebra. If a polygon is given in a figure, then instead of measuring the height and the base of each triangle separately, it will be more convenient first to transform the polygon into one triangle. This is done in the following way – each triangle is separated by a line, passing through the ends of two adjacent sides, is replaced by other equal sides, which will decrease the number of its sides by one, when it is connected to the other parts of the polygon. For that it is necessary that the associated triangle keeps its base and hight separately, that one of its sides is shared with the polygon, and also some of its two other sides form a straight line with the adjacent side of the polygon. All of this should have been well known to my students and for this reason I am not discussing it further.

Let us now talk about the right trapezoid and parallelogram. *Right trapezoid* is called quadrangle, in which three sides are perpendicular to one another. In the quadrangle $ABDC$ (Fig. 29), if three sides a, b, c are perpendicular to one another, but AC is not perpendicular to BA and hence to CD, then the perpendicular from point A to the opposite side CD should be equal to c and cut off from CD on the side of c a line, equal to a, which is already implied by the inequality of the sides b and a. Let $b > a$, then the perpendicular c from A to b will go inside the trapezoid and will cut off from one side a rectangular, whose area $= ac$, on the other side of the right triangle, whose sides are c and $b - a$, the area $= \frac{1}{2}s(b-a)$. Two areas give $\frac{1}{2}c(a+b)$, then the area of the right trapezoid) is equal to the half of the product of the middle perpendicular and the sum of the two last.

Parallelogram is called quadrangle, made of two equal triangles, connected in such a way that their equal sides are opposite. The base of the parallelogram is called one of its sides, the hight of the triangle for the same reason is called the hight of the base. So the area of the parallelogram is equal to the product of the hight end the base.

8 On Parallelograms

Two lines are called *parallel* when they lie on the same plane and do not intersect, regardless of their extension.

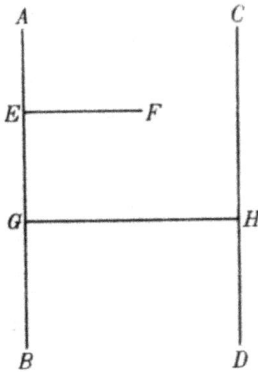

Fig. 30

If from a point on a line we draw a perpendicular to a line that is parallel to the first, then the perpendicular will be perpendicular to the second line as well. Otherwise a perpendicular to one line would form an astute angle with the perpendicular to the other line and hence the given parallel lines would converge on one or the other sides of the perpendicular. It follows from here, that through a given point not lying on a given line we can draw only one parallel to that line; it can be made parallel when it is perpendicular to the perpendicular on the given line from the given point.

A perpendicular to one of two parallel lines, being extended on one or the other side, will meet the other parallel line and hence it will be also perpendicular to it. If AB is parallel to CD, and EF perpendicular to AB (Fig. 30), then from some point H on the line CD we draw perpendicular GH to AB. Four lines are formed: CH, GH, GE, EF, perpendicular to one another, because the last EF should fall under right angle to the first CH. When HG and EF are on the opposite side of the line AB, then the continuation of EF on the other side should meet CD, as we mentioned it above. Finally, if point G happens to be in E, then perpendiculars EF

and *GH* will be forming one line.

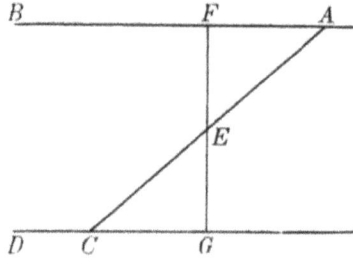

Fig. 31

Two parallel lines, which are intersected by a third line, give a sum of two angles on the other side of the third line, facing one another, and are equal to two straight angles. Let *AB* and *CD* – are two parallel lines, *AC* intersects them (Fig. 31). Let divide *AC* at the point *E* into two equal parts, after we draw from *E* a perpendicular to *AB*, continue it on the other side until intersects *DC*. Let this perpendicular is represented by the line *FEG*. Then we get two equal triangles *AEF* and *CEG*, because they have two angle and one side are equal. The remaining two angles *BAE* and *GCE* are also equal, hence, $\angle BAC + \angle DCA = 200°$.

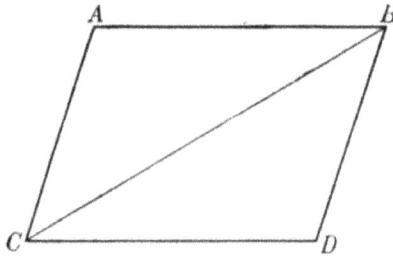

Fig. 32

On the contrary, if we assume that $\angle BAC + \angle DCA = 200°$ and only *EF* is perpendicular to *AB*, then in $\triangle AEF$ and $\triangle GEC$ we will have again two angles equal on the side, therefore the third will be equal to third, that is, *EG* is a perpendicular to *GC* and the two lines *AB* and *CD* are parallel. When the sum of the angles $BAC + DCA = 200°$, then the angles at the intersection point of the lines *AC* and *CD* are equal.

A parallelogram, by definition, is composed by two equal triangles, because the diagonal in the parallelogram forms with the opposite sides equal angles at the intersection point, and therefore the opposite sides of the parallelogram are parallel. In the parallelogram *ABCD* (Fig. 32) the two triangles, comprising it, *ABC*, *BCD* are equal, $\angle ABC = \angle BCD$, hence *AB* is parallel to *CD*. By the same reason *AC* is also parallel to *BD*. From here we see, that the property of parallelograms is that (1) opposite sides are equal, (2) opposite angles are equal and (3) opposite sides are parallel.

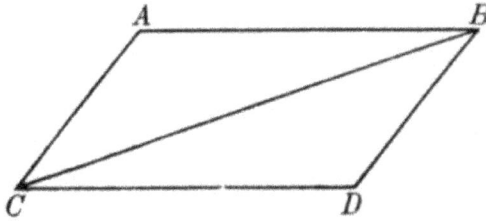

Fig. 33

Assume that in the square $ABCD$ the opposite sides are equal (Fig. 33). Draw a diagonal BC, and we get two equal triangles, because they have three equal sides, and therefore quadrangle $ABCD$ is a parallelogram.

A quadrangle, which has only two equal and parallel opposite sides will be parallelogram. Let in $ABCD$ the side $AB = CD$ and AB and CD are parallel, then the angle $ABC = \angle BCD$, $\triangle ABC$ is equal to $\triangle BCD$, hence $ABCD$ is a parallelogram.

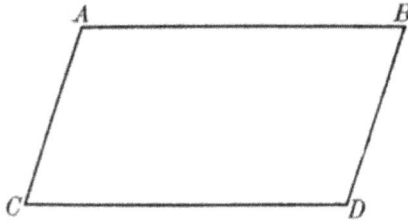

Fig. 34

If in the quadrangle the opposite sides are parallel, then the quadrangle will be parallelogram. If AB is parallel to CD, AC to BD, then the angle $ABC = \angle BCD$, $\angle ACB = \angle CBD$, $\triangle ABC$ is equal to the triangle BCD, therefore $ABCD$ is a parallelogram.

If in the quadrangle the opposite sides are equal, then it will be a parallelogram (Fig. 34). If in the quadrangle $ABCD$ the angle $A = \angle D$, $\angle B = \angle C$, then $ABCD$ is a parallelogram. It is possible to prove that the sum of the angles in each quadrangle is $= 400°$, hence in $ABCD$ the sum $\angle A + \angle C = \angle B + \angle D = 200°$, that is, the lines AB and CD are parallel. Also we can prove, that $\angle A + \angle B = 200°$, and therefore AC is parallel to BD.

In order to prove that in every quadrangle the sum of the angles $= 400°$, we should prove that in each triangle the sum of the angles is $= 200°$. In the rectangle each angle is right so the sum of all angles is $= 400°$. In the rectangular triangle, because two of them form a rectangle, the sum of the angles $= 200°$. We saw, that every triangle can be formed from the sum or the difference of two right triangles. In the triangle ABC (Fig. 35) when the perpendicular AD to BC passes through the middle point,

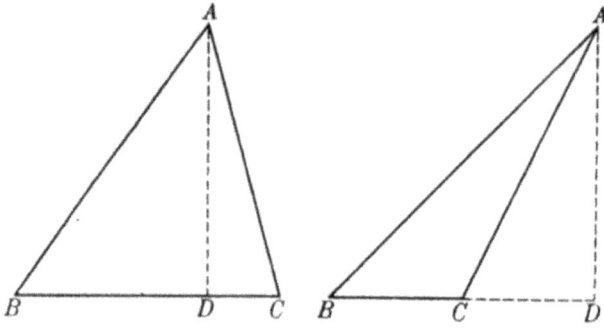

Fig. 35

$\angle B + \angle BAD = 100°$, $\angle C + \angle CAD = 100°$, hence all three angles $\triangle ABC = 200°$. In triangle ABC, when perpendicular AD to the BC [is situated] out of the triangle, then the $\angle B + \angle BAD = 100°$, $\angle ACD + \angle CAD = 100°$, from where $\angle B + \angle BAC = ACD$, that is $\angle A + \angle B + \angle ACB = 200°$.

9 ON MEASURING PRISMS

Two planes are called *parallel* when they do not intersect, regardless of how much they are extended. From this definition it follows that a plane intersecting two parallel planes gives as a result of the intersection two parallel lines.

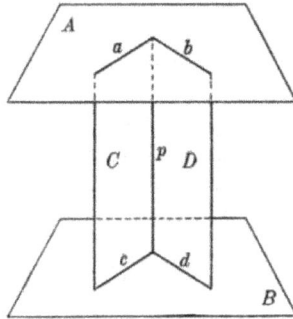

Fig. 36

Let us imagine two planes A and B which are parallel (Fig. 36). Then we take two converging lines a and b situated on one of the planes and through them we draw two planes C and D, perpendicular to B. We will have three lines: p – a line of intersection of C and D, perpendicular to B, the line c from the intersection of C and B and the line d from the intersection of D with B. Lines a and c are parallel, the lines b and d are also parallel, p is perpendicular to c and d, therefore p is perpendicular to a and b. Hence p is perpendicular to the plane A. So the perpendicular to one of the two parallel planes will be perpendicular to the other as well.

From here we see that it is possible to draw through a given point a parallel plane to the given. Namely, when we draw a perpendicular from a given point on a given surface and after that to this perpendicular draw a perpendicular plane through the given point. From the definition of parallel planes it is seen that through a given point on a given plane we can draw only one parallel plane, otherwise their intersection with the new plane would give two lines through one point, which would be parallel to one line only.

A perpendicular to a plane if extended in both directions should meet the parallel plane, and therefore would be a perpendicular to it.

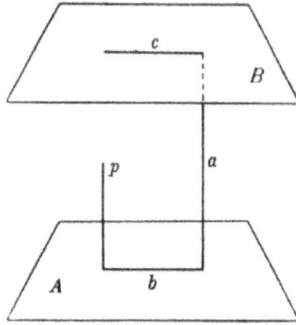

Fig. 37

Imagine plane A, and then another plane B parallel to A (Fig. 37). Let the line p is perpendicular to A. Take an arbitrary point on the plane B and draw from it a perpendicular a to the plane A. After that through a and p draw a plane, whose intersection with A gives the line b, and the intersection with B – the line c. In such a way we will have four lines c, a, b, p in one plane and perpendicular to one another, so the first line c should meet the last p.

The magnitude of the perpendiculars between parallel planes is everywhere equal, because two perpendiculars with lines, drawn between them in the planes, form a rectangle.

A body, bounded by planes intersecting in parallel lines, and two parallel planes intersecting all first planes, is called a *prism*. Two figures, whose planes are parallel and intersect all others, will be the *bases* of the prism, and the perpendicular between the two bases is called the *hight*, and the parallel lines connecting the bases are called *edges*.

All planes bounding the prism, with the exception of the base, are rectangles with opposite parallel sides, and therefore are parallelograms. They are called *sides* of the prism. The number of the sides is equal to the number of the edges of the base. By the number of sides, prisms are three-sided, four-sided and so on.

In a three-sided prism the bases are equal, because the two triangles, which are the bases of the prism, the three sides are equal. Each prism can be divided into three-sided prisms by planes, passing through two of the edges of two adjacent sides, and hence in each prism the bases are equal.

Two three-sided prisms are equal, when they have equal bases, one of the sides, the angle of this side with the base and the hight. The proof is self-evident.

It follows from here, that when the base of the prism is a parallelogram, then a plane through the corresponding diagonals on the base divides the prism into two three-sided equal prisms. In the prism P (Fig. 38), if the bases $ABCD$, $A'B'C'D'$ are parallelograms, then the plane $ACC'A'$, passing through two corresponding diagonals AC and $A'C'$, divides prism P into two parts, whose bases are equal. Further, if we draw a perpendicular p from AC to $A'C'$ and draw a from the end of p perpendicular to AC

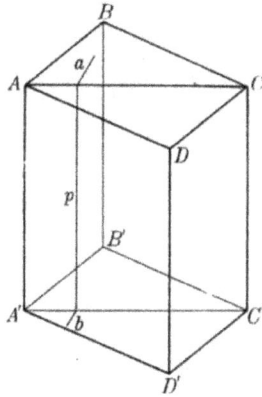

Fig. 38

in the plane $ABCD$, we imagine a plane through p and a, then we get a line of intersection b, perpendicular to $A'C'$ and parallel to a. Hence the angle of the plane ABC with the plane $ACC'A'$ is equal to the angle of the plane $A'C'D'$ with plane $ACC'A'$, because the parallelogram $ACC'A'$ is common to the two prisms, in which P is divided. Their hight is also common and as a result they are equal.

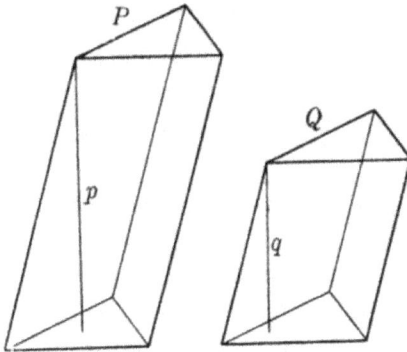

Fig. 39

Two three-sided prisms, which have equal bases, one of the side is equal and the angle of inclination of this side to the base, but the heights are different. If in two prisms P and Q (Fig. 39) the bases are equal, but the heights are presented as p and q, then we divide the first height into p parts, and the second in q parts: then all parts will be equal. We draw, through points of intersection, planes, which are perpendicular to the height, then the prism P will be divided into p equal prisms and equal to q prisms in which prism Q will be divided by planes, perpendicular to its height at the points of intersection. Therefore the content of prism P to the content of prism Q will be the same as the content of the heights p and q.

When the prism's edges are perpendicular to the base, then the prism is called *right prism*. Two right prisms, which have equal heights, and the bases are rectangles which the bases of the prisms. Let the given two right prisms A and B, with one base – a rectangle with sides p and q, and the second – rectangle with sides n and m, then dividing sides p, q, n, m correspondingly into p, q, n, m parts, in the point of the intersection draw perpendiculars, and through the perpendiculars – planes, which are perpendicular to the base, then the prism A will be divided into pq prisms, prism B into nm prisms, which will be equal, hence the prisms A and B are contained as bases.

For measuring of bodies we take for a unit a cube, whose edges are linear units. Comparing with such a cube a right prism P, whose hight is unit, and its base – rectangle A, understanding A as the area of the rectangle, we see that the quantity of prism P will be A. Comparing with the prism P with another prism Q, whose base is the same, but the hight is p. Then we see that quantity [of the prism] Q is equal to the product of the quantities of the prism P and the hight p. So it is equal to pA, then it is the product of the hight and the base. So the right prism, whose base is a rectangle, is equal of the product of the hight and the base.

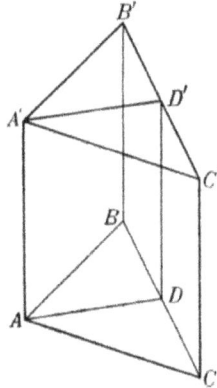

Fig. 40

In the right prism, whose base is a rectangle, the planes through the corresponding diagonals on the base form two equal prisms, whose bases are right triangles and can have as edges two arbitrary lines. From here it follows that the quantity of the right prism, whose base is a right triangle is equal to the product of the hight.

In the three-sided right prism P (Fig. 40), any of the triangles ABC and $A'B'C'$ had not been a base, it always should find such a side BC, at which both angles will be acute. Draw from the opposite angle A a perpendicular AD to BC and draw a plane through AD and AA'. Such a plane crosses the side $B'C'BC$ in the line DD' perpendicular to the base ABC, so it divides the prism P into two right prisms, whose bases are the right triangles ABD and ACD. Let p is the hight of prism P, M is the area of the triangle ABD, N is the area of the triangle ADC, then the quantities of two prisms forming P, will be pM and pN, and the quantity of both is $p(M+N)$, that is, the quantity of the three-sided prism is equal to the product of the base and the hight.

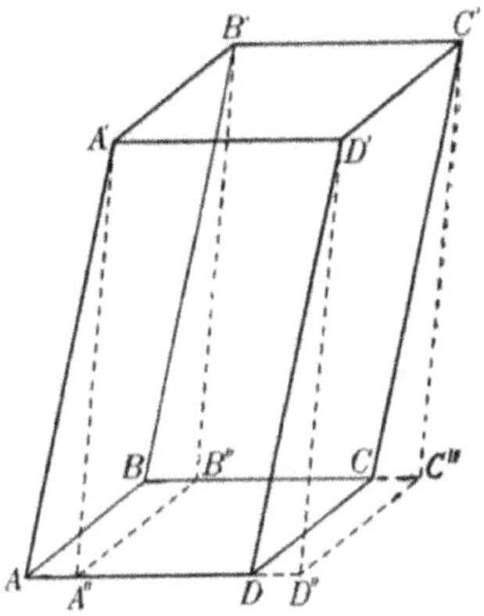

Fig. 41

In an [inclined] prism AC' we draw the bases $ABCD$, $A'B'C'D'$ which are parallelograms (Fig. 41). From A' and D' draw perpendiculars to AD, from B', C' to BC. One of the perpendiculars, for example $A'A''$, should lie in the prism; then $B'B''$ will also lie in the prism, and the other two $C'C''$ and $D'D''$ will be outside of the prism. Connect points A'' and B'' with the points C'' and D'' by lines. If $B'B''$ is the line of intersection of the plane through $A'B'$ and $A'A''$, then $B'B''$ will be perpendicular to BC; but because two perpendiculars from B' to BC could not be drawn, it is the same that we suggest that $B'B''$ is a line of intersection of the plane through $A'B'$ and $A'A''$, or a perpendicular to BC. So the points A', B', A'', B'' lie in the same plane as the points C', D', C'', D''.

Prisms, one of which cuts off the plane $A'A''B''B'$, and the other connects to the plane $D'D''C''C'$, are equal, because their bases $ABB'A'$ and $DCC'D'$ are equal, $\triangle AA''A'$ is equal to the $\triangle DD''D'$ and the angles between the sides and the bases. So, the prism between the parallelograms $ABCD$ and $A'B'C'D'$ is equal to the prism between the bases $A''B''C''D''$ and $A'B'C'D'$.

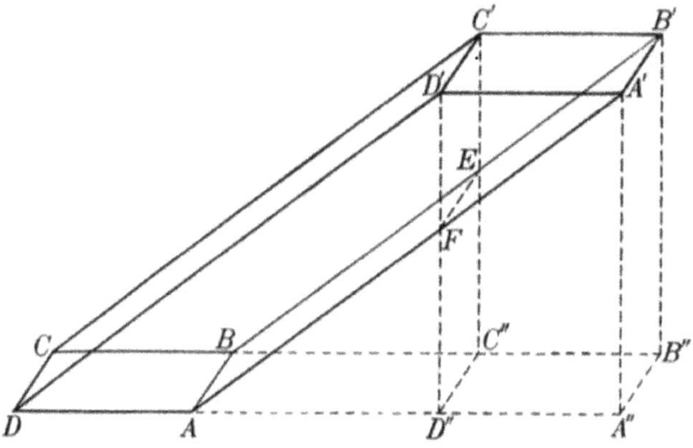

It could happen (Fig. 42) that both points A'' and D'', as well as B'',

Fig. 42

C'', will be outside of the prism; then we will have a proof also for the equality of the prisms $ABB''A''A'B'B$ and $DCC''D''D'C'C$, from which when we take away their common part the prism $ABC''D''FEB$, we get such equal bodies, which when connected with the prism $A'D'FEC'B'$ give two equal prisms: one between the bases $ABCD$ and $A'B'C'D'$, the other between $A''B''C''D''$ and $A'B'C'D'$, as in the first case. On the base of the above we conclude, that the quantity of two prisms is equal, if the bases are equal parallelograms and are situated between two parallel planes and if some of their opposite sides also are placed between two parallel planes. Therefore, taking into account one quantity, if the sides of the prism, whose base is a parallelogram, are not perpendicular to the base, we can substitute the opposite sides by others, perpendicular to the base, only that the hight rests the same. Hence each prism, whose base is a parallelogram, is equal to a right prism with the same base and the same hight.

As it was proved, a prism, whose base is a parallelogram, is divided into two three-sided prisms by equal planes, passing through the diagonals of the base, and as sides of the parallelogram and its diagonal can be taken completely arbitrary, then it follows from here, that every three-sided prism is equal by quantity to another, which has the same base and hight. The right three-sided prism is equal to the product of the hight and the base, therefore any three-sided prism is equal to the product of hight and base.

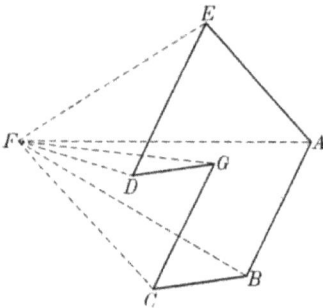
Fig. 43

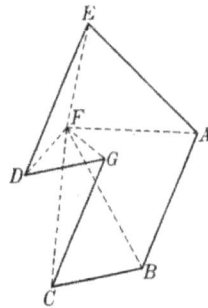
Fig. 44

A prism, whose base is a polygon could be divided in three-sided prisms. For this it is needed only to divide the base into triangles, after that to divide the other base into respective equal triangles, and draw planes through the lines of dividing. Dividing polygons into triangles has already been explained, but here we will give another method. In order that the polygon $ABCGDE$ be divided into triangles we need to take a point F in the plane the polygon inside, outside of it or on its circle, then from point F we draw in all angles the lines a, b, c, g, d, e. In such a way there will be formed triangles from the sides of the polygons and the lines, drawn to F, whose number will be equal to the sides of the polygon $ABCGDE$.

Starting with that of the triangles, which lies fully or partially on the plane of the polygon, it is necessary to add the next triangle where their angles at F will be added and subtract each time a triangle, as soon as the angle at F is subtracted and it will continue till that moment when all triangles are considered. Such a connection of triangles gives the area of a polygon. For example (Fig. 43), when point F is outside the polygon $ABCGDE$, the area of the polygon $ABCGDE = \triangle AFE + \triangle AFB + \triangle BFC - \triangle CFG + \triangle GFD - \triangle DFE$. The area of the polygon $ABCGDE$ is equal to $\triangle AFB + \triangle BFC - \triangle CFG + \triangle GFD + \triangle DFE + \triangle EFA$ when F is taken inside the polygon (Fig. 44).

Let now, generally speaking, the area O of the base of the prism P is formed from the union of the triangles A, B, C, D and so on, so that $O = A + B + C + D +$ and so on, where some of the triangles should be considered negative in their union for the formation of the area O. Imagine a plane through point F, from which lines are drawn, dividing the base O into triangles, and through the prism's edges, then all such planes will intersect in one line, passing through F, and will divide prism P into three-sided prisms A', B', C', D' and so on, whose bases will respectively be A, B, C, D and so on. Considering those of them negative, whose bases inter with the negative sign in the sum $- A + B + C +$ and so on. Evidently, the prism $P = A' + B' + C' + D'$ and so on. Let us call p the hight of the prism P. It will be together with the hight and three-sided prisms $A', B', C'D'$ and so on. Therefore $A' = pA, B' = pB, C' = pC, D' = pD$, and the quantity of the prism $P = p\{A + B + C + D +$ and so on$\} = p.O$, is the product of the hight and the base.

10 MEASUREMENT OF PYRAMIDS AND ALL BODIES BOUNDED BY PLANES

A *pyramid* is a body, of which one of the bounded planes can be a triangle or polygon and is called *base*, the other planes – triangles, coming out from one point which is called *top of the pyramid*.

A perpendicular drawn from the top of the pyramid to its base is called *hight*. Triangles, connecting the top and the base of the pyramid are called – *sides*; their lines of connection – *edges*.

Pyramid, where one of its edges is perpendicular to the base and therefore serves as its hight, will be a *right pyramid*.

By the number of the sides the pyramid is three-sided, four-sided and so on.

If we draw a plane through the hight and an edge, then the base of the pyramid will be divided into triangles, and the pyramid itself will be divided into three-sided right pyramids, from which the given pyramid is composed through addition. It is obvious from here that for the measurement of the pyramid we need to know only the volume of each three-sided pyramid.

We divide the right three-sided pyramid P of hight $DC = p$ into two equal parts at point G (Fig. 45); then we draw a perpendicular plane to the hight DC; it will create intersection points F and E with the other two edges AF, BE; as a result we will get the triangle FEG and the three-sided right pyramid $DFGE$. Through the side EF we draw a plane, which is perpendicular to the base ABC of the pyramid P: then we will get K and I as points of intersection with the sides AC and BC, and the lines FK, EI, KI. These last three lines together with the lines FE, FG, EG, KC, IC form a prism, whose base is the triangle KCI, the hight is the line GC, equal to $\frac{1}{2}p$. Hence its magnitude is $= \frac{1}{2} \triangle KCI.p$.

We draw a line IH through the point I on the base ABC to the side AB, parallel to AC, we connect the points of H and E through the line EH. The parallelograms $FEHA$, $FEIK$, $KIHA$ form a prism with the triangles FKA, EIH, which can be considered as half of a prism with FK is the hight, the parallelogram $AKIH$ is its base, and hence its volume is equal to half of the product of the area of the parallelogram $AKIN$ and the height $\frac{p}{2}$.

47

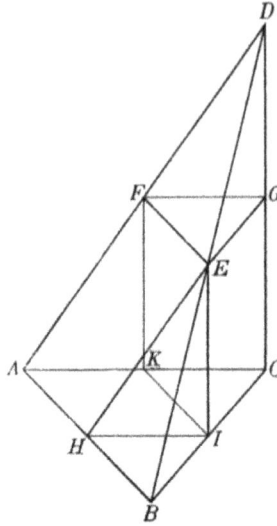

Fig. 45

Triangles DEG, EBI are equal, because $DG = EI$; the angles at points D and E are equal; the angles at the points G and I are right; therefore, $DE = BE = \frac{1}{2}BD$; $GE = BI = \frac{1}{2}BC$. As $FG = AK = \frac{1}{2}AC$, $AF = FD = \frac{1}{2}AD$. The triangles FDE, HEB are equal, because $DE = EB$, $FD = HE$, the angles at points D and E are equal, so $FE = HB = \frac{1}{2}AB$. The triangles FGE, HIB are equal, because their sides are equal, and as $EI = DG$, then the pyramids $HIBE$, $FGED$ are equal, which can be easily verified by placing one over the other.

The magnitude of the pyramid $FEGD$ will be denoted by the letter P_1. The triangles KCI, HIB are equal and are equal to the half of the parallelogram $AKIH$, hence, the area of the base ABC, denoted by the letter O, will be the area of the parallelogram $AKIH = \frac{1}{2}O$; the area of the triangle $KCI = \frac{1}{4}O$. The magnitude of the pyramid P = prism $FGEIKC$+prism$AFKIHE$+2pyramid$FEGD = \frac{1}{4}p \cdot O + 2P_1$. For the base of the pyramid P_1 we can accept the triangle FGE, i.e., $\frac{1}{4}O$, the height – the line DG, i.e. $\frac{1}{2}p$. So if denote by P_2 the magnitude of such a pyramid, whose base is $\frac{1}{4^2}O$, the height $\frac{1}{2^2}p$; through P_3, whose base is $\frac{1}{4^3}O$, hight $\frac{1}{2^3}p$; through P_4, whose base $\frac{1}{2^4}p$, in general through P_n the magnitude of such a pyramid, whose base $\frac{1}{4^n}O$, the hight $\frac{1}{2^n}p$, where n is the whole positive number, then the magnitude of the pyramids P, P_1, P_2, P_3 and so on till P_n can be defined with the help of each other from such equations:

$$P = \frac{1}{4}pO + 2P_1;$$

$$P_1 = \frac{1}{4}\frac{1}{2}p\frac{1}{4}O + 2P_2;$$

$$P_2 = \frac{1}{4}\frac{1}{2^2}p\frac{1}{4^2}O + 2P_3;$$

$$P_3 = \frac{1}{4}\frac{1}{2^3}p\frac{1}{4^3}O + 2P_4;$$

...

$$P_{n-1} = \frac{1}{4}\frac{1}{2^{n-1}}p\frac{1}{4^{n-1}}O + 2P_n.$$

Inserting in the first equation, instead of P_1, its value from the second equation; after, instead of P_2, its value from the third equation and so on to the last P_{n-1} then we find that

$$P = \frac{1}{4}pO\left\{1 + \frac{1}{4} + \frac{1}{4^2} + \frac{1}{4^3} + \ldots + \frac{1}{4^{n-1}}\right\} + 2^n P_n$$

$$= \frac{1}{3}pO - \frac{1}{3.4^n} + 2^n P_n.$$

If instead of P_n, we take the magnitude of a prism whose base is $\frac{1}{4^n}O$, the height $\frac{1}{2^n}p$, then such a prism will be greater than P_n and therefore $P - \frac{1}{3}pO < \frac{2}{3.4^n}pO$. Regardless of the kind of the numbers p and O we always can find such a number n, that $\frac{2}{3.4^n}pO$ will be less of any given number, and therefore the difference $P - \frac{1}{3}pO$ should be less then any number, that is, $P = \frac{1}{3}pO$.

If a pyramid is not right and three-sided, then we can divide it by planes, passing through the height, into three-sided right pyramids A', B', C' and so on, whose bases the triangles A, B, C and so on and whose magnitudes $\frac{1}{3}pA, \frac{1}{3}pB, \frac{1}{3}pC$, and so on, and because the connection of the pyramids A', B', C' and so on for forming of P is the the same which as what is needed for the connection of the triangles A, B, C and so on in order to form the base of the pyramid P then, from here it follows, that the magnitude of each pyramid is equal to one third of the prodict of the base and the hight.

Each body, bounded by polygons, can be formed by pyramids. For that reason it is important to take a point wherever outside of the body, on its surface or inside, and then draw planes through it and the polygons of the body. In such a way, we can form pyramids whose number is equal to the number of polygon.

11 MEASUREMENTS OF THE CIRCUMFERENCE AND THE AREA OF THE CIRCLE

The method of measuring of curved lines, planes, bounded by curved lines, curved surfaces and bodies bounded by such surfaces, is, as we already explained, to divide the curved lines and surfaces into very small parts, replacing them by straight lines or planes, and then to look for such a limit to which is approached by the quantity[1] of a curved line, surface or body as the parts of division are becoming smaller. Such a limit is accepted for the sought quantity. In real measurements, i.e with the help of chains, threads in general elastic bodies, also try to be closer to such a quantity.

In order to define the quantity of straight lines, which are replacing very small parts of a curve, and from here to find the quantity of the curve, it is important to know how to find the quantity of one side of the triangle, when its given parts correspond to the conditions of equality. But for measuring circles, spheres and any geometrical quantities, where we have such lines and surfaces, it is necessary to know only how one of the sides of a right triangle is defined with the help of two adjacent.

If in two triangles all angles are equal, then their sides, and angles are proportional. Assume that (Fig. 46) angle $A = \angle A'$, $\angle B = B'$, $\angle C = \angle C'$; further we assume that the content of the sides a to a' is expressed by two whole numbers p and q. We divide side a in p parts and side a' respectively in q parts. These and the other parts will be equal. Draw through points of division parallel lines to side b in the triangle ABC and lines parallel to b' in the other triangle. We will draw between these lines through the dividing points in one triangle lines parallel to c and in another parallel to c'. For example let side a be divided into three parts, lines ef, $e'g$ drawn parallel to b and lines ef', $e'f''$ parallel to c, then the formed here triangles Bef, $ee'f'$, $e'Cf''$ will be equal, because they have one equal side and the angles are equal, so $Bf = ef' = e'f''$, $ef = e'f' = Cf''$. From here it is seen, that the side c is divided by parallel lines into three equal parts. Regardless of the dividing of side a, we always will have that the

---

[1]EDITOR'S NOTE: By the word "величина" (translated here as "quantity" or "magnitude") Lobachevsky means "length", "area" or "volume."

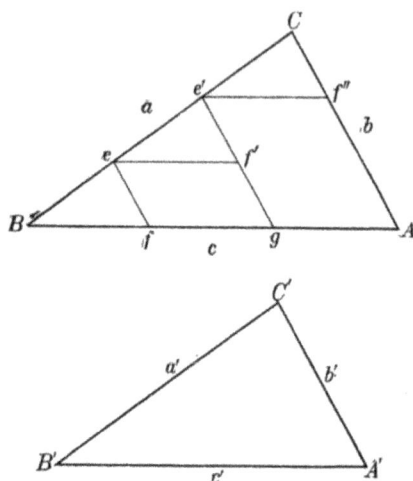

Fig. 46

lines parallel to side b, drawn through points of division, will divide side c into such equal parts, as the number of parts of side a. The same will happen in the triangles $A'B'C'$, where the parts of side c' will be equal the parts of side c, because they belong to the equal triangles. So if a is divided into p parts, a' into q parts, which parts are equal, then side c is also divided into p parts, side c' into q parts, i.e. the content of side c to c' remains the same. It is obvious that we can prove the same for b and b'. Such triangles as ABC and $A'B'C'$ are called in Geometry *similar*.

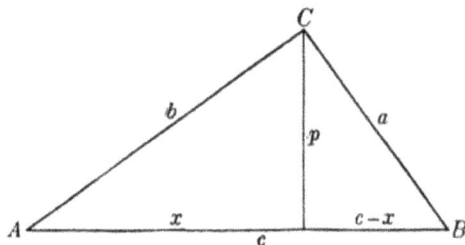

Fig. 47

In the right triangle ABC (Fig. 47) from the right angle C we draw a perpendicular p to the hypotenuse c. The part of the hypotenuse adjacent to angle A will be called x, the other will be $c-x$, the side opposite to angle B will be denoted by b, the other side $-a$. The perpendicular p divides the right triangle into two right triangles, which are similar to triangle ABC. It follows from here, that $x = \frac{b^2}{c}$, $c-x = \frac{a^2}{c}$. Combining the two equations we find $c^2 = a^2 + b^2$, that is, the square of the hypotenuse is equal to the sum of the squares of the sides. Knowing the two sides of the right triangle we can find the third side.

Assume now that a radius r of an arc is known and its chord c is also known. From here we can find the quantity of the chord c' half of the

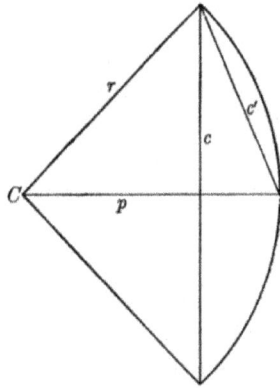

Fig. 48

given arc. First we determine the perpendicular p from the center of the chord c, and after we find it $p = \sqrt{r^2 - \frac{c^2}{4}}$. In the right triangle (Fig. 48), whose hypotenuse c', sides $\frac{1}{2}c$ and $r - p$ we find

$$c' = \sqrt{\frac{1}{4}c^2 + (r - p)^2},$$

when replacing p we get,

$$c' = r.\sqrt{2 - \sqrt{4 - \left(\frac{c}{r}\right)^2}}.$$

A chord of the arc 200^0 has diameter $2r$ for that, because the chord of the arc:

$$\frac{200}{2} = r.\sqrt{2}$$

$$\frac{200}{2^2} = r.\sqrt{2 - \sqrt{2}}$$

$$\frac{200}{2^3} = r\sqrt{2 - \sqrt{2 + \sqrt{2}}}$$

$$\frac{200}{2^4} = r.\sqrt{2 - \sqrt{2 + \sqrt{2 + \sqrt{2}}}}$$

$$\frac{200}{2^5} = r.\sqrt{2 - \sqrt{2 + \sqrt{2 + \sqrt{2 + \sqrt{2}}}}}$$

From here it is seen that chord of the arc

$$\frac{200}{2^n} = r.\sqrt{2 - \sqrt{2 + \sqrt{2 + \sqrt{2 + \ldots}}}}$$

where the number 2 under the square is multiplied n times. The number, which is multiplied by r, should be as smaller as the number n is greater, and, following the rule for measuring curved lines, the number, of which we are talking about, being multiplied by 2^n, will give an arc in 200^0 and it will be more likely that n will increase. In such a way we can find that an arc of 200^0 is equal to

$$r.3,141592653\ldots$$

A number, by which r is multiplied here, can be found only through approximation of decimal fractions. It is usually denoted in the mathematical books by π, so the quantity of the arc of 200^0 will be expressed by πr. The quantity of each other arc, which is of q degrees, will be $\frac{q}{200^0}\,\pi r$; from here we see that the arcs of equal degrees are contained as diameters.

We have still to show that by the number π we understand the limit to which the product 2^n and the chord of the arc $\frac{200}{2^n}$ approaches is greater when the number n is greater.

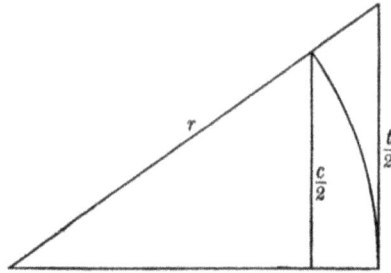

Fig. 49

Let us call c a chord whose arc is $\frac{200^0}{2^n}$, then we draw a radius through the middle of the chord c to the circumference of the circle and here we draw a perpendicular (Fig. 49) whose quantity is between the radius passing through the ends of the chord c and call it t. Find:

$$t = \frac{c.r}{\sqrt{r^2 - \frac{1}{4}c^2}}$$

from here comes the difference

$$t - c = c\frac{r - \sqrt{r^2 - \frac{1}{4}c^2}}{\sqrt{r^2 - \frac{1}{4}c^2}}$$

The number $r - \sqrt{r^2 - \frac{1}{4}c^2}$ denotes the distance between the lines c and t, which is as smaller as n is greater, and can be done smaller than any given number. Further calling K the area of the circle of radius r we see that

$$K < 2^{n-1}.tr$$

and

$$K > 2^{n-1}.cr$$

the difference $2^{n-1}.r(t-c)$, as it represents area between lines c and t, is as smaller as n is greater, and maybe smaller for each quantity; so $2^{n-1}c.r$ and it is as closer to K as the greater n is. So it is possible that π is found through approximation and the area of the circle

$$K = \pi.r^2$$

12 Of Measuring the Volume of Cylinder and Cone, the Surfaces of Right Cylinder and Right Cone

A *cylinder* is called a body, bounded by two parallel equal circles and curved surface whose property is such that it is intersected into right lines by planes, passing through the center of the circles. From this definition it is clear, that the cylinder is a prism whose base is a circle. The line, connecting the centers of the circles of the cylinder, is called *axis*, and each of the circles – a *base*, the perpendicular between the bases is called *hight*. Cylinders are *right*, when its axis is perpendicular to the base and *inclined* in the opposite case.

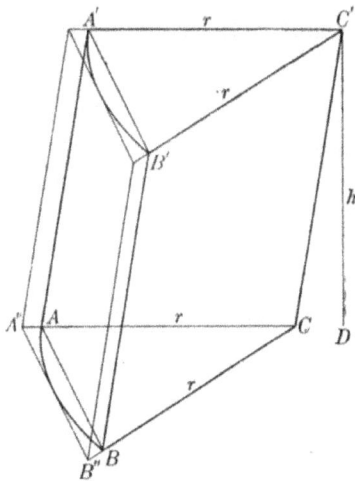

Fig. 50

Imagine the circumference of the base of a cylinder is divided into 2^n parts, where n is a whole number. Imagine planes going through the points of division and the axis of the cylinder. Then the whole cylinder will be divided into 2^n parts of such a kind, as we see in the figure (Fig. 50). The line CC' is the axis of the cylinder, lines $AC, BC, A'C', B'C'$ are equal to the radius of the base, which will be denoted by r, the curved lines

57

$AB, A'B'$ are 2^n-th parts of the circumference of the base, the line $C'D$ is the hight of the cylinder, which will be denoted by h. If a plane is drawn through the lines AA', BB', then the intersection of this plane with the segments of the circles ACB, $A'C'B'$ gives straight lines AB, $A'B'$, which will be chords of the arcs $\frac{400^0}{2^n}$; and the quantity of the formed here prism $A'C$ will be the product of the triangle ABC and its hight h. The quantity of all similar prisms around the axis CC' of the cylinder will be $2^n \cdot h \cdot ABC$. This quantity will represent the quantity of the cylinder as closely as the number n will be greater. Because the real quantity of the cylinder is bounded between $2^n \cdot h \cdot ABC$ and $2^n \cdot h \cdot A''CB''$, where $A''CB''$ is the area of the triangle, which is bounded by the extensions of radii AC, BC and the line $A''B''$, perpendicular to the radius r in the middle of the arc AB. But the difference $2^n \cdot A''CB'' - 2^n \cdot ABC$, as we saw, becomes as smaller as n is getting greater, and it can be made even smaller than any arbitrary number; hence $2^n \cdot h \cdot ABC$ is as closer to the real quantity of the cylinder as n is greater. The product $2^n \cdot ABC$ is representing the base of the cylinder as closely as n is greater; so the real quantity (volume) of the cylinder is the product of the base and the hight as the quantity of any prism is equal to $\pi r^2 h$.

In order to find the magnitude of the curved surface of the cylinder, it is necessary instead of its parts $ABB'A'$ take parallelograms $ABB'A'$ and connect their areas; but these parallelograms will be equal only if the cylinder is right. In all cases, it is easy to see, that parallelograms $ABB'A'$ will be a rectangular, whose area = $h.AB$, and therefore the sum of all parallelograms on the surface of the cylinder = $2^n \cdot AB \cdot h$, whose quantity represents the real quantity of the surface of the cylinder as closely as n is greater. But we saw that $2^n \cdot AB$ is as closer to the quantity of the circumference of the circle, whose radius is r, as n is greater. So the real quantity of the curved surface of the right cylinder is equal to the product of the hight of the cylinder and the circumference of the base, that is, $2\pi \cdot r \cdot h$.

A *cone* is called a body bounded by a circle and a curved surface with such a property that there is a point on it throught which and the center of the circle we draw planes and they intersect the curved surface of the cone into straight lines. Such a definition shows that a cone is a pyramid whose base is a circle. The circle of the cone is called a *base*; a point through which a straight light line can be drawn to the circumference of the base of the cone's surface, is called *the top*, and the line connecting that point to the center of the base – *axis*, the perpendicular from the top to the base is called *hight*. Cones are two types – *right*, when their axis play the role of the hight, otherwise the cones are called *inclined*.

Imagine that the base of the cone is divided into 2^n equal parts and through the points of division and the axis of the cone are drawn planes. These planes divide the cone into 2^n such parts, as we see in the figure (Fig. 51). Point D is the top of the cone, C is a center of the base, the lines AC, BC are radii of the base, which let be denoted by r, the line

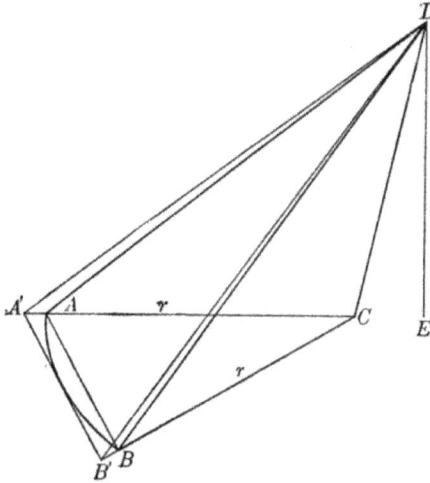

Fig. 51

DC is the axis of the cone, the line $DE = h$ is the hight of the cone, the curved line AB is 2^n-th part of the circumference of the base. Let us draw a plane through points D, A, B, the quantity of the pyramid $DABC$ will be $\frac{1}{3}h \cdot ABC$, and the quantities of all such pyramids in the cone around axis DC will be $\frac{1}{3}h \cdot 2^n ABC$. Let us extend the radii AC and BC until they meet $A'B'$, drawn perpendicular to r in the middle of the arc AB. The real quantity of the cone is between $\frac{1}{3}h \cdot 2^n ABC$ and $\frac{1}{3}h \cdot 2^n A'CB'$. The difference $2^n ACB - 2^n A'CB'$ is as smaller as n is greater, and can be done smaller than any number. On the other hand, the product $2^n \cdot ABC$ approaches the magnitude of the circle of radius r as closer as the number n is greater; hence the real quantity (volume) of the cone is equal to the product of the base and one third of the hight, as in the case of the pyramid, that is, $= \frac{1}{3}h\pi r^2$.

In the right cone triangles ADB are equal, because instead of the parts of the curved surface ADB, we put triangle ADB and multiply by 2^n, we will get the quantity of the curved surface of the cone as correctly as n is greater. The area $\triangle ADB = \frac{1}{2}AB\sqrt{AD^2 - \frac{1}{4}AB^2}$; the line AD is equal in all triangles, and there is a distance between the top of the cone from the circumference of the base; it can be found with the help of the hight h of the right cone and the radius r of the base, namely $AD = \sqrt{h^2 + r^2}$. So the surface of the cone $= \frac{1}{2} \cdot 2^n \cdot AB\sqrt{h^2 + r^2 - \frac{1}{4}AB^2}$, which expression must be taken as a limit; but the limit of the product $2^n AB$ is the circumstance of the base $2\pi r$, and the line AB approaches as closer to zero as n is greater; so the surface of the right cone $= \pi \cdot r\sqrt{h^2 + r^2}$.

13 ON THE MAGNITUDE OF THE VOLUME AND THE SURFACE OF THE SPHERE

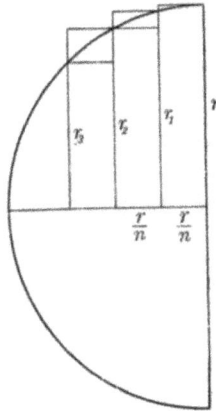

Fig. 52

Let us divide the sphere into two equal parts by a plane passing through its center. Draw parallel planes to this plane, which are at equal distances from one another, and in such a way divide the radius r of the sphere into n parts, where n is a whole number. Each half of the sphere will be divided in such a way into n parts by parallel planes, whose intersections will be circles. Considering only one half of the sphere and intersecting it again with a plane, passing through the radius, divided into n parts, the intersection of this plane with half of the sphere will look like what is depicted in the figure (Fig. 52). Denote by r_1, r_2, r_3 and so on the radii of the circles which are situated from the center of the sphere at $\frac{r}{n}, \frac{2r}{n}, \frac{3r}{n}$ and so on. If we take the sum of n of the cylinders, whose hights are equal to $\frac{r}{n}$, and the radii of the base are r, r_1, r_2, r_3 and so on till r_{n-1}, it is clear, that such a sum will be greater than half of the volume of the sphere. On the contrary, if we take sum $n-1$ of cylinders, removing the first, then such a sum will be less than half of the volume of the sphere, and so the true magnitude of the sphere will be between:

$$2\pi\frac{r}{n}\{r^2 + r_1^2 + r_2^2 + \ldots + r_{n-1}^2\}$$

and

$$2\pi\frac{r}{n}\{r_1^2 + r_2^2 + r_3^2 + \ldots + r_{n-1}^2\}$$

The difference between two limits is $\frac{\pi r^3}{n}$ and so can be done smaller than any given number by increasing n. From here it is seen, that the volume of the sphere is getting as closer to one of the two mentioned here limits, as the number n is greater. The radii r_1, r_2, r_3, r_4 and so on are defined by r and n in the following way;

$$r_1^2 = r^2 - \frac{r^2}{n^2}; \quad r_2^2 = r^2\left(1 - \frac{2^2}{n^2}\right); \quad r_3^2 = r^2\left(1 - \frac{3^2}{n^2}\right) \text{ and so on}$$

$$r_{n-1}^2 = r^2\left(1 - \frac{(n-1)^2}{n^2}\right),$$

then we get the magnitude of the volume of the sphere:

$$2\pi r^3\left\{1 - \frac{1}{n^3}(1 + 2^2 + 3^2 + \ldots + (n-1)^2)\right\}.$$

Instead of the sum of the squares, which here we multiply by $\frac{1}{n^3}$ and write as $\frac{n}{6}(2n^2 - 3n + 1)$;; it can be verified by setting $n = 1, = 2, = 3$; in order that it holds for each whole number n, it is necessary that

$$\frac{n}{6}(2n^2 - 3n + 1) + n^2 = \frac{n+1}{6}\{2(n+1)^2 - 3(n+1) + 1\},$$

which is indeed correct. So, the volume of the sphere approaches as closer to

$$2\pi r^3\left\{\frac{2}{3} + \frac{1}{2n} - \frac{1}{6n^2}\right\}.$$

as n is greater, and since these parts where n is in the denominator goes as closer to zero as n is greater; hence the true magnitude of the volume of the sphere $= \frac{4}{3}\pi r^3$.

The method of determining the magnitude of the volume of the sphere, proposed here, can serve for determining the magnitude of part of the sphere cut off by a plane. Let h is a distance between two parallel planes, and one of which passes through the center; let the magnitude of the length h to radius of the sphere r be two whole numbers m and n. The numbers m and n can be multiplied by equal numbers and hence the number n can be increased infinitely. Dividing the distance h into m equal parts and drawing through the point of division parallel planes to the two given, the part of the sphere between the two outer planes will be as closer to

$$\pi r^3\left\{\frac{m}{n} - \frac{1}{n^3}(1 + 2^2 + 3^2 + \ldots + (m-1)^2)\right\},$$

as the number n is greater. The sum of the squares, which is multiplied by $\frac{1}{n^3}$, as we saw, $= \frac{1}{3}m^3 - \frac{1}{2}m^2 + \frac{1}{6}m$. Hence $= \pi r^3 \left\{ \frac{m}{n} - \frac{1}{3}\left(\frac{m}{n}\right)^3 + \frac{1}{2n}\left(\frac{m}{n}\right)^2 - \frac{1}{6n^2}\left(\frac{m}{n}\right) \right\} = \pi r^3 \left\{ \frac{h}{r} - \frac{1}{3}\left(\frac{h}{r}\right)^3 + \frac{1}{2n}\left(\frac{h}{r}\right)^2 - \frac{1}{6n^2}\left(\frac{h}{r}\right) \right\}$.

The terms, which can be dived by n, approach as closer to zero as the number n is getting greater; so the true magnitude of the parts of the sphere between two planes $= \pi\left(r^2 - \frac{1}{3}h^2\right)h$. Taking out this magnitude from the volume of half of the sphere, we get the magnitude of the segment $\frac{1}{3}\pi.(2r^3 - 3r^2.h + h^3)$, or put $r-p$ instead of h, the magnitude of the segment will be $\frac{1}{3}\pi.p^2(3r-p)$. If we add to this segment a cone, whose hight is h, the distance between the top of the circumference to the base r, then we will get a conical segment of the sphere which is $\frac{2}{3}\pi r^2 p$. Difference of such a canonical segment with other, where line p will be $p-a$, we find $\frac{2}{3}\pi r^2.a$, and so the magnitude of such segment of the sphere does not depend on the location of the line a to the center.

In order to find the magnitude of the surface, following the general rule, we should divide it into extremely small parts and place together such parts of the plane, which, being connected, will give the magnitude of the surface as accurate as the parts are smaller. It is simpler to divide the surface of the sphere first by planes around the radius, and after then by planes perpendicular to the radius. As a result we will have parts of the surface, bounded by four arcs, and then it is easy to see, that four points of intersecting the arcs lie in one plane. Because, connecting the four points with straight lines, we need to take a part of the plane between the four lines instead of the parts of the curved surface. Starting connecting the parts between two subsequent parallel planes, as far as here the number of the parts does not depend on the parts of radius, then assuming that it increases limitlessly, the whole part of the surface situated between the two planes will approach to the parts of the conical surface, drawn between the circles, bounded by parallel surface, which will therefore be the true magnitude of this part of the surface of the sphere.

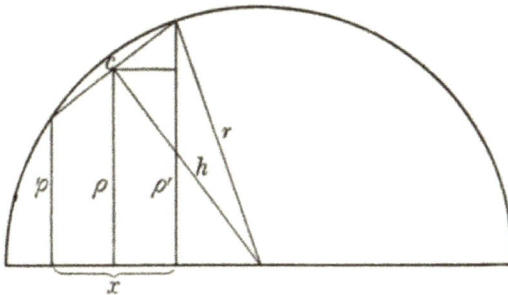

Fig. 53

Let us call r the radius of the sphere, c – the distance between the circumferences of two parallel subsequent planes ; $'\rho$ - the radius of the smaller circle, ρ' – the radius of the greater circle, the magnitude of the conical surface between two citcle will be $\pi.c('\rho + \rho')$. Imagine that half of the sphere is separated by a plane through a radius to which $'\rho$ and ρ' are perpendicular. After that we intersect this half by a plane, passing through the same radius (Fig. 53). On the last plane we draw the lines $'\rho$ and ρ' and the chord c. Then we draw a perpendicular ρ from the middle of the chord c to the radius and we connect the middle point of the chord c by a line h with the center of the sphere. We call x the distance between $'\rho$ and ρ'. If from the middle of c we draw a perpendicular to ρ' we get a right triangle, whose hypotenuse is $\frac{1}{2}c$, the sides are $\frac{1}{2}x$ and $\rho'-\rho$ similar to the right triangle, whose hypotenuse is h, the sides are ρ and the distance x to the center. From here we find $c = x.\frac{h}{\rho}$. The sum $'\rho + \rho'$ is equal to 2ρ, so $\pi c('\rho + \rho') = 2\pi.h.x$. For simplicity let chords c between the parallel planes are equal; then the lines h for all c will be also equal, and the sum of all canonical surfaces from one end of the radius to the other will be $4\pi r.h$, which will depict the surface of the sphere as closer as the chord c will be smaller; but if chord c is smaller, then h is closer to r, hence the true magnitude of the surface of the sphere is $= 4\pi r^2$, that is, it is equal to four big circles of the sphere.

It is possible to find the surface of the sphere in this way too: regardless of the division of the surface into parts, we put instead of them planes s and draw perpendicular $r - p$ to s, it is necessary that the sum of all products $\frac{1}{3}s(r - p)$ would be as closer to $\frac{4}{3}\pi r^3$ as s is smaller. But as the plane s is smaller as p is closer to zero, and the sum of the plancs s – closer to the surface of the sphere, which should be $4\pi r^2$, as found above.

GEOMETRICAL INVESTIGATIONS ON THE THEORY OF PARALLEL LINES
1840

INTRODUCTION

I have found some problems in geometry, which I consider to be the reason why this science, because it did not evolved to use calculus, has not advanced even a single step beyond its state since the time of Euclid. The main part of these problems I think is the lack of clarity in the fundamental notions of geometrical quantities, the methods for measuring these quantities, and at the end the important gap in the theory of parallel lines, all efforts of today's mathematicians to explain it have been unsuccessful. The efforts of Legendre did not added to this theory anything new, because he had been forced to abandon his research, and to use theories, ideas, which he tries to show as necessary axioms.

My first attempt on the foundations of geometry I published in the *Kazan newspaper* in 1829. Hoping that I have satisfied all requirements, I started later on to work additionally on this science and this work of mine I published in several parts in "The learned notes of Kazan university" in 1836, 1837 and 1838 entitled "New foundations of geometry with a complete theory of parallel lines." The quantity of my work may prevent my compatriots from following the subject, which after Legendre had lost its appeal. I think the theory of parallel lines should not have been neglected by mathematicians; for this reason I intend to present here the essence of my research on them; and I would like to mention that, despite the opinion of Legendre all other imperfections, as for example the definition of a straight line, turns out to be irrelevant and lacking any influence on the theory of parallel lines.

1 PRELIMINARY PROPOSALS

In order to make the book more easy to understand I am giving in advance some important definitions and explanations.

1) *Straight line is covering itself in all locations.* By this I mean that when rotating a surface the straight line does not change its place, if it passes through two stationary points of the surface.

2) Two straight lines cannot intersect in two points.

3) A straight line, being enough extended, in both directions, should go beyond any limits and in such a way divides a bounded plane into two parts.

4) Two straight lines, perpendicular to the same third straight line, never intersect, regardless of their extension.

5) A straight line always intersects another straight line, if it goes from one side of the line to the other.

6) Vertical angles, are equal when the side of one angle is continuation of the other. It holds for both flat right angle and flat two-sides angles.

7) Two straight lines cannot intersect, if some other third right line intersects them under equal angles.

8) In the right triangle the equal angles lie against equal sides and vice versa.

9) In the right triangle the greatest side lies against the greatest angle. In the right triangle the hypotenuse is greater then each of the sides, and the adjacent to it angles are astute .

10) The right triangles are congruent, if they have equal sides and two angles or two sides and the angle between them, or two sides and an angle lying against the the greater side, or three sides.

11) A straight line, perpendicular to two other straight lines, which do not lie in the same plane, is perpendicular to all straight lines, drawn through the point of their common intersection in the plane of the two last straight lines.

12) The intersection of a sphere by a plane is a circle.

13) A straight line, which is perpendicular to the line of intersection of two perpendicular planes and lies in one of these planes, is perpendicular to the other plane.

14) In a spherical triangle equal sides are facing equal angles, and vice versa.

15) Spherical triangles are congruent, if they have two equal sides and an angle between them, or a side with adjacent angles.

Starting from here, our further suggestions will be explained and proved.

2 PARALLEL LINES

16) All straight lines originating from one point on a plane, can be divided in two classes with respect to a given straight line on the same plane, namely, *intersecting* it and not *intersecting* it. The *boundary line* of both classes of these straight lines is called *parallel to the given line*.

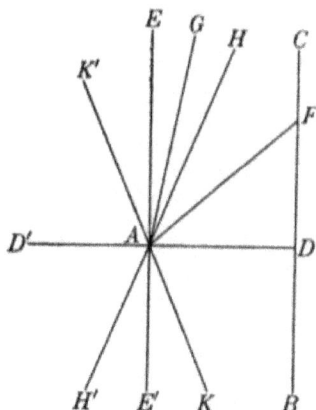

Fig. 1

From point A (Fig. 1) we draw to the given line BC a perpendicular AD, to which, in turn, draw a perpendicular AE. In the right angle EAD the straight lines coming out from point A, either all intersect line DC, as for example AF, or some, as perpendicular AE, do not meet the line DC. Not knowing whether perpendicular AE is the only line, which does not intersect DC, we will consider that it is possible, that there exist some other lines, for example AG, which does not intersect DC, regardless of how much we extend them. In the transition from the intersecting line AF to the non-intersecting line AG we must meet line AH, parallel to DC, – the bordering line, – on one side where no line AG do not intersect DC, by the way when on the other side each line AF intersects line DC. The angle HAD between parallel AH and perpendicular AD is called *the angle of parallels* (the of angle parallelism); we will denote it here by $\Pi(p)$ at when $AD = p$.

If $\Pi(p)$ is a right angle, then the extension AE' of the perpendicular

71

AE will also be a parallel extension of line DB of the line DC. We should notice here, that regarding the four right angles, which at the point A form perpendiculars AE and AD and their extensions AE' and AD', each straight line, coming out from point A, either itself or through its extension is located in one of these two right angles, which face line BC, so that all other straight lines if sufficiently extended in both directions, except parallel EE', should intersect line BC.

If $\Pi(p) < \frac{1}{2}\pi$, then on the other side of the perpendicular AD, under the same angle $DAK = \Pi(p)$, one more line AK parallel to the extension DB of the line DC can be drawn. In such a way, while assuming this we have to distinguish also the *side of parallelism*. The remaining lines or their extensions inside the two right angles, facing BC, belong to the intersecting lines, if they lie inside the angle $HAK = 2\Pi(p)$ between the parallels. On the contrary, they belong to the non-intersecting AG, if they are situated on the other side of the parallel AH and AK in the opening of the two angles

$$EAH = \frac{1}{2}\pi - \Pi(p), \quad E'AK = \frac{1}{2}\pi - \Pi(p)$$

between parallels and perpendicular EE' to AD. In a similar way on the other side of the perpendicular EE' the extensions AH' and AK' of the parallels AH and AK will also be parallel to BC. The remaining lines belong to the angle $K'AH'$ to the intersecting, and in the angles $K'AE$ and $H'AE'$ – to the non-intersecting.

With this in mind, assuming $\Pi(p) = \frac{1}{2}\pi$ lines can be only intersecting or parallel; if we assume that $\Pi(p) < \frac{1}{2}\pi$, then we should assume two parallels, one on the one side of the perpendicular, the other on the other side of it; besides, between the remaining lines we should distinguish intersecting and non-intersecting lines. In both suggestions, the indication (property) of parallelism is that a line becomes intersecting at the smallest deviation to that side, where the parallel line lies. In such a way, if AH parallel to DC, then every line AF, regardless of at how small the angle HAF is intersects DC.

17) *The straight line keeps the property of parallelism at of all its points.*

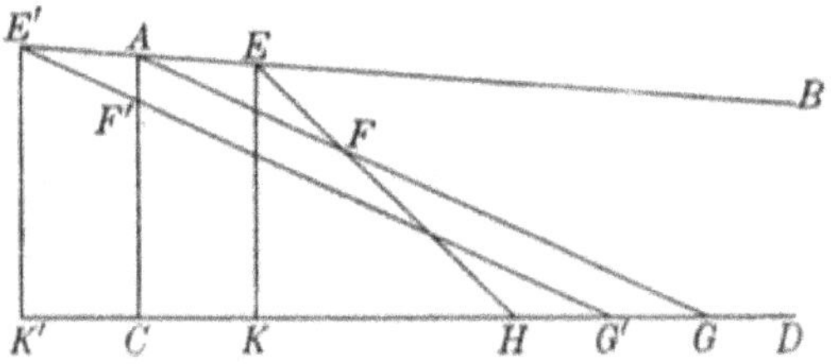

Fig. 2

Let the straight line AB (Fig. 2) be parallel to CD [at the point A] and let AC be perpendicular to the latter. We consider two points, which are taken arbitrarily: one on the line AB and the other on its extension on the other side of the perpendicular.

Let assume that point E is on this side of the perpendicular, with which AB is considered as parallel to CD. From point E we draw a perpendicular EK to CD, after we draw EF in such a way that it passes inside the angle BEK. The points A and F will be connected by a straight line, whose extension should meet CD somewhere at G (Proposal 16). Then we get the triangle ACG, inside of which the line EF goes. As the latter line cannot intersect AC due to the mere construction, also cannot meet either AG or AK (Proposal 2), then it should meet CD at some point H (Proposal 3).

Let now E' be a point on the extension of line AB and let $E'K'$ be a perpendicular to the extension of line CD; let us draw a line $E'F'$ under such a small angle $AE'F'$, in order that it intersects AC somewhere at F'; after, from A we draw again under the same angle with AB the line AF whose extension will meet CD at G (Proposal 16). In such a way we get a triangle AGC, where the extension of lines $E'F'$ goes; as that line cannot secondarily cross AE, and also cannot intersect AG, because the angle $BAG = BE'G'$ (Proposal 7) then it should meet CD somewhere at G'.

So regardless from which points E and E' [on the straight line AB] the lines EF and $E'F'$ originated and how little they deviated from line AB, they will always intersect the line CD which is parallel to AB.

18) *Two lines are always mutually parallel.*

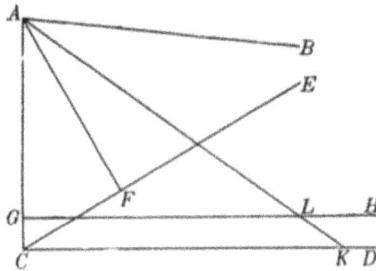

Fig. 3

Let AC be a perpendicular to line CD (Fig. 3), to which AB is parallel; we draw from C a line CE under any astute angle ECD to CD and from A draw a perpendicular AF to CE. Then we will get the right triangle ACF, where the hypotenuse AC is greater than the side AF (Proposal 9). Put $AG = AF$ and put AF on AG; then AB and FE will take the place of AK and GH, where the angle $BAK = FAC$. So, AK should cross the line DC anywhere at point K (Proposal 16). In such a way we get the triangle AKC, inside of which the perpendicular GH goes. It meets the line AK at L (Proposal 3) and defines on the line AB the distance AL the points of intersection of the lines AB and CE from point A.

From the above it follows, that CE will always meet AB, regardless of how small the angle ECD is. For this reason CD is parallel to AB

(Proposal (16).

3 THE SUM OF THE INNER ANGLES OF THE RIGHT TRIANGLE

19) *In a right triangle the sum of the three angles cannot exceed two right angles.*

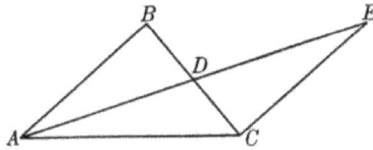

Fig. 4

Assume that in the triangle ABC (Fig. 4) the sum of the three angles is equal to $\pi + \alpha$; if its sides are not equal, we take the smallest of them BC, divide it into two equal parts at point D, draw from A through D a line AD and its extension DE is equal to AD. After we connect point E with point C by a straight line EC. In the equal triangles ADB and CDE the angle $ABD = DCE$ and $BAD = DEC$ (Proposals 6 and 10). From here it follows, that in the triangle ACE the sum of angles should be also equal to $\pi + \alpha$; except this, the smallest angle BAC of the triangle ABC (Proposal 9) transfers into a new triangle ACE, where it is divided into two parts EAC and AEC. Continue in the same way, each time dividing into two equal parts the side, which is opposite the smallest angle, we arrive at a triangle, where the sum of its three angles is equal to $\pi + \alpha$, but there we will find two angles, each of which by absolute magnitude is smaller then $\frac{1}{2}\alpha$; as the third angle cannot be greater than π, then α should be either zero or negative.

20) *If in any right triangle the sum of its three angles is equal to two right, then it holds for any other triangle.*

Assume that in the right triangle ABC (Fig. 5) the sum of the three angles $= \pi$; in such a case two of its angle should be astute.

Fig. 5

75

From the top B of the third angle we draw to the opposite side a perpendicular p. Then the triangle ABC will be divided into two right triangles. In every of the triangles the sum of the three angles should also be equal to π, because the sum cannot be greater then π in either of them, and in the formed from them triangle the sum of the angles should not be smaller then π. In such a way we get a right triangle with sides p and q,

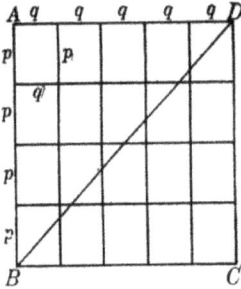

Fig. 6

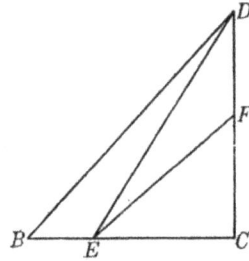

Fig. 7

and from it we form a parallelepiped, in which the opposite sides are equal, and the adjacent sides to p and q are mutually perpendicular (Fig. 6). Applying for a second time the same parallelepiped, we can form a similar parallelepiped with sides np and q, and, finally, the parallelepiped $ABCD$ with mutually perpendicular sides, where $AB = np$ $AD = mq$, $DC = np$, $BC = mq$, where m and n are arbitrary whole numbers. Such a parallelepiped is divided by the diagonal BD into two equal right triangles BAD and BCD, in each of which the sum of the three angles is equal to π. The numbers m and n can be chosen in such way that the right triangle BCD (Fig. 7), whose sides are $CD = np$, $BC = mq$, enclosed another given [right] triangle EFC and as long as their right angles will be superimposed. If we draw a line DE, then we get more right triangles, from which every two consecutively have a common side. The triangle BCD is formed by connecting two triangles BDE and DEC. In neither of them the sum of the three angles can be greater then π; it should be equal to π, because in the formed triangle that sum must be equal to π. We observe that triangle EDC contains two triangles DEF and CFE; hence in the triangle CEF the sum of the three angles should be equal to π. In general, this should hold for every triangle, because all of them can be divided into two right triangles.

From here it follows, that only two assumptions are possible: either the sum of three angles in all right triangles is equal to π or the sum is smaller then π in all triangles.

21) *From a given point can always be drawn a straight line in such a way, that it forms with the given straight line an arbitrarily small angle.*

From a given point A (Fig. 8) we draw to a given straight line BC a perpendicular AB; we take an arbitrary point D on BC and draw line AD; further, we make $DE = AD$ and draw AE. Let in the right triangle

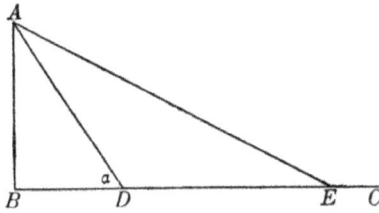

Fig. 8

ABD the angle $ADB = \alpha$. In such a case in the isosceles triangle ADE the angle AED should be either equal to $\frac{1}{2}\alpha$, or smaller (Proposals 8 and 19). Continuing in such a way, we finally come to such an angle AEB, which is smaller then any given angle.

22) *If two perpendiculars to the same straight line are also parallel, then in the rectilinear triangles the sum of the three angles is equal to* π.

Assume that the lines AB and CD (Fig. 9) are mutually parallel and are perpendicular to AC. From A we draw lines AE and AF towards the points E and F, taken on the line CD, at any distances $FC > EC$ from point C. Suppose that, in the right triangle ACE the sum of the three angles is equal to $\pi - \alpha$, and in the triangle AEF it is equal to $\pi - \beta$. In such a case the triangle ACF [the sum of the angles] will be equal to $\pi - \alpha - \beta$, and nether α nor β can be negative. Let further the angle $BAF = a$, $AFC = b$; in such a case $\alpha + \beta = a - b$. Now taking away the line AF from the perpendicular AC, we can form an angle a between AF and the line AB extremely small; also we can decrease angle b as well. Therefore, two angles α and β cannot have any other magnitude except $\alpha = 0$ and $\beta = 0$.

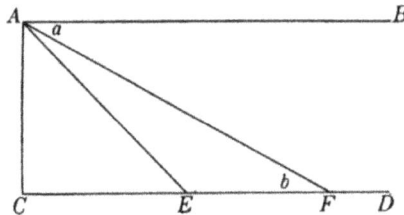

Fig. 9

From here it follows, that in all rectilinear triangles the sum of the three angles either equal to π and then the angle of parallelism is $\Pi(p) = \frac{1}{2}\pi$ for any line p, or in all triangles this sum is $< \pi$, and then also $\Pi(p) < \frac{1}{2}\pi$.

The first assumption serves as a base of the *ordinary geometry* and *plane trigonometry*. The second assumption can also be considered since it does not lead to any contradiction in the results. It justifies a new geometrical theory, which I call *imaginary geometry* and which I intend to present here and develop up to deriving equations about the sides and angles of right and spherical triangles.

4 INVESTIGATION OF THE ANGLE OF PARALLELISM

23) *For each given angle α it can be found such a line p in order that* $\Pi(p) = \alpha$.

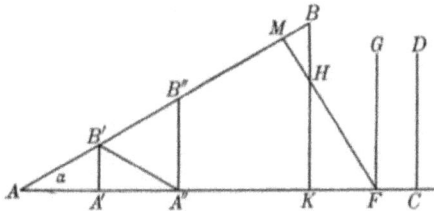

Fig. 10

Let AB and AC (Fig. 10) are two straight lines whose intersection forms an astute angle α. Let take an arbitrary point B' on AB and from this point we draw a perpendicular $B'A'$ to AC and make $A'A'' = AA'$, draw from A'' a perpendicular $A''B''$ and continue in such a way until we reach a perpendicular CD, which already does not meet AB. This should be so because if in the triangle $AA'B'$ the sum of all three angles is equal to $\pi - \alpha$, then in the triangle $AB'A''$ it is equal to $\pi - 2\alpha$, in the triangle $AA''B''$ it is smaller then $\pi - 2\alpha$ (Proposal 19) and so on, until it, at the end, becomes negative and in such a way demonstrates that a triangle cannot be formed. The perpendicular CD can be namely that to which all perpendiculars from points, lying near A, intersect AB; in any case from the transition from intersecting to non-intersecting such a perpendicular must exist. Now from point F we draw a line FH forming with FG an astute angle HFG, and namely on this side on which point A lies. From any point H on the line FH we draw to AC a perpendicular HK, whose extension, therefore, should intersect AB somewhere at B. It forms in such a was a triangle AKB, in which the extension of the line FH goes, and therefore it must meet the hypotenuse AB somewhere at M. As GFH is an arbitrary angle and can be taken to be very small, then the line FG is parallel to AB and $AF = p$ (Proposals 16 and 18).

It is easy to see, that with the decreasing of p the angle α increases, approaches at $p = 0$ to $\frac{1}{2}\pi$; with the increasing of p the angle α is getting

79

smaller, and approaches to zero when $p = \infty$. As it is completely arbitrarily, which angle to consider by the symbol $\Pi(p)$, when the line p is expressed by a negative number, then we accept

$$\Pi(p) + \Pi(-p) = \pi$$

which equation should hold for all values of p as positive as negative and also for $p = 0$.

5 Mutual Position of Parallel Lines

24) *The farther parallel lines extend in the direction of parallelism the faster they approach one another.*

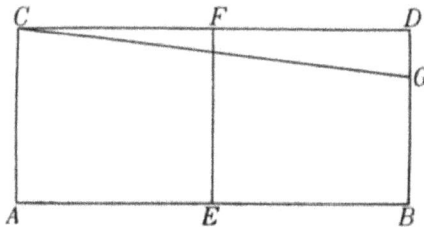

Fig. 11

We draw two perpendiculars $AC = BD$ to the straight line AB (Fig. 11) and connect their end points with a straight line [CD]. Then the quadrangle $CABD$ will have two right angles at A and B, and at C and D – two astute angles (Proposal 22), which are equal – this can be easily seen by imposing this quadrangle on itself, in order that line BD falls on AC and AC – on BD. Divide AB into two equal parts and at the point of division E we draw a perpendicular EF to AB. It should be also perpendicular to CD, because quadrangles $CAEF$ and $FEBD$ covered each other, if we impose them on each other so that line FE stays in the same position. As a result of this the line CD cannot be parallel to AB, the parallel to the last line at point C, namely CG, should be inclined to AB (Proposal 16) and to cuts off from the perpendicular BD a part $BG < CA$. As point C is chosen on the line CG arbitrarily, from here it follows that CG approaches AB as closer as far as we extend it.

25) *If two straight lines are parallel to a third line they are also parallel to each other.*

Let us first suppose that three lines AB, CD, EF (Fig. 12) lie on the same plane. If two of them AB and CD are parallel to the other line EF, then AB and CD are also parallel. In order to see it, draw a perpendicular AE from an arbitrary point A on the outer line AB, a perpendicular AE to other outer FE, which will intersect the middle line CD at some point C (Proposal 3) under an angle $DCE < \frac{1}{2}\pi$ with the side of the parallel CD

81

82

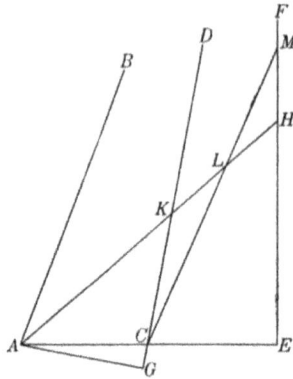
Fig. 12

to EF (Proposal 22). The perpendicular AG drawn from the same point A to CD must go inside the opening of the astute angle ACG (Proposal 9). Any other line AH, drawn from A inside the angle BAC, should cross the line EF, which is parallel to AB, somewhere at H, regardless of how small the angle BAH may be. Therefore, CD will intersect the line AH in triangle AEH at K, because the line cannot meet with EF. If AH passed through point A inside the angle CAG, then the line should be in the triangle CAG intersecting the extension CD somewhere between points C and G. From here it follows that AB and CD are parallel (Proposals 16 and 18).

If we accept, that both outer lines AB and EF are parallel to the middle line CD, then every line AK, drawn from point A inside the angle BAE, will intersect the line CD at point K, regardless of how small the angle BAK may be. On the extension of AK we take an arbitrary point L and connect it with C by the line CL, which intersects EF at M. As a result the triangle MCE is formed. The extension of line AL inside the triangle MCE cannot intersect for a second time either AC or CM. Therefore, it should meet EF somewhere at H. For that reason AB and EF are mutually parallel.

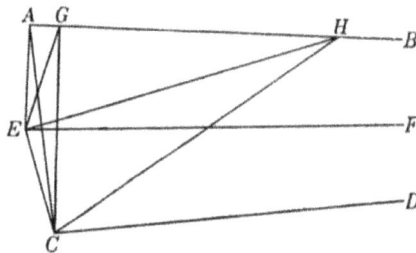
Fig. 13

Let now the parallel lines AB and CD (Fig. 13) lie in two planes, whose intersection is the line EF. From an arbitrary point E on EF we draw a perpendicular EA to one of the two parallel lines, for example to

AB. After, from the base A of the perpendicular EA we draw again a perpendicular AC to the second parallel line CD and then connect the ends E and C of both perpendicular by the line EC. The angle BAC should be astute (Proposal 22); hence, the perpendicular CG, drawn from C to AB, falls at point G on that side of CAQ where we consider lines AB and CD parallel. Each line EH, regardless of how little it is deviated from EF, lies with EC on a plane, which should intersect the plane of the two parallel lines AB and CD along some line CH. This last line intersects somewhere the line AB, namely at point H, which belongs to all three planes and through which the line EH should pass as well; so EF is parallel to AB. In a similar way can find the parallelism between two lines EF and CD.

In accordance with this assumption, that the line EF is parallel to one of the two other parallel lines AB and CD, means, that EF is considered as intersection of two planes in which the two parallel lines AB and CD lie. Hence, two lines are parallel, if they are parallel to the same third line, even though they lie in different planes. The latter suggestion can be expressed also in the following way: *three planes intersect into lines, all of which are parallel, because it is supposed that two of the lines are parallel.*

6 MEASURING TRIHEDRAL ANGLES

26) *Opposite triangles on the surface of a sphere have equal areas.*

Under opposite triangles we mean such triangles, which are formed by the intersection of the surface of a sphere [the same planes] on the both sides of the center. In such triangles the sides and the angles have opposite directions.

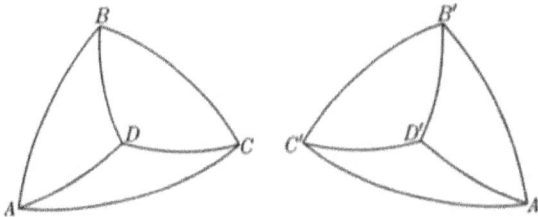

Fig. 14

In the opposite triangles ABC and $A'B'C'$ (Fig. 14, where one of the triangles should be presented as a mirror image) the sides $AB = A'B'$, $BC = B'C'$ $CA = C'A'$ and their angles at the points A, B, C are also equal to the corresponding angles in the other triangle at points A', B', C'. Imagine a plane, passing through three points A, B, C, and a perpendicular, drawn to this plane from the center of the sphere. The extension of this perpendicular in both sides will intersect the opposite triangles at points D and D' of the sphere's surface. The distance between point D from points A, B, C on the sphere in the arcs of the big circles should be equal (Proposal 12) and the distance $D'A'$, $D'B'$, $D'C'$ should be equal in the other triangle (Proposal 6). Hence the isosceles triangles are arranged around points D and D' in both spherical triangles ABC and $A'B'C'$ are congruent.

In order to explain the equality of two surfaces, consider the following suggestion: *two surfaces are equal if they are formed by addition or subtraction of two equal parts.*

27) *Trihedral solid angle is equal to the half-sum of its twohedral angles without the right one.*

In the spherical triangle ABC (Fig. 15), where each side $< \pi$, we will denote the angles with A, B, C, extend the side AB thus that a complete

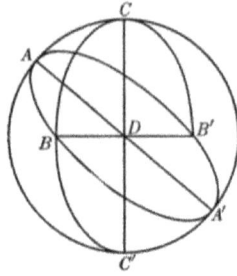

Fig. 15

circle $ABA'B'A$ is formed, which divides the sphere into two equal parts. In the half of the sphere where the triangle ABC is situated, we extend two other sides beyond the point of their intersection C till their intersection with circle at points A' and B'. This semi-sphere will break into four triangles ABC, ACB', $B'CA'$, $A'CB$ whose quantities are P, X, Y, Z. It is clear, that here

$$P + X = B,$$

$$P + Z = A$$

The quantity of spherical triangle Y is equal to the quantity of its opposite triangle ABC', where the side AB is common with the triangle P, the third angle C' lies at the end point of the diameter of the sphere, coming from C through the sphere's center D (Proposal 26). From here it follows, that $P + Y = C$; as $P + X + Y + Z = \pi$ then we get also

$$P = \frac{1}{2}(A + B + C - \pi).$$

To this result we can come through a different way, considering only the suggestion, given above regarding the relative equality of areas. (Proposal 26).

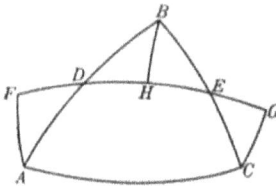

Fig. 16

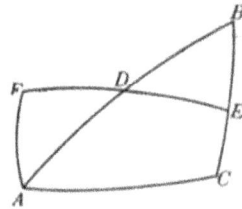

Fig. 17

In the spherical triangle ABC (Fig. 16) we divide the sides AB and BC in two equal parts and through the points of division D and E we draw a big circle; from points A, B, C, we draw to this circle perpendiculars AF, BH, CG. If the perpendicular from B goes in H between D and E, then the triangle BDH formed in this way is equal to AFD, and BHE is equal to EGC (Proposals 6 and 15). From here it follows that the area of the triangle ABC is equal to the area of quadrangle $AFGC$ (Proposal 26). If point H coincides with the middle point E of the side BC (Fig. 17), then only two equal right triangles AFD and BDE are formed, replacing one by the other, we prove the equality of the areas of the triangle ABC and the quadrangle $AFEC$. If, at the end, point H goes outside of triangle ABC (Fig. 18), then we move from the triangle ABC to the quadrangle $AFGC$, adding triangle $FAD = DBH$ and then subtracting the triangle $CGE = EBH$. If we imagine in the spherical quadrangle $AFGC$ big circles,

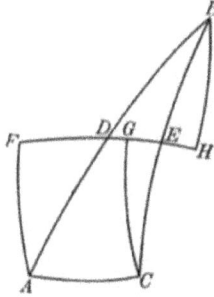
Fig. 18

passing through points A and G and through F and C, then the arcs between AG and FC are equal (Proposal 15); so that the triangles FAC and ACG are congruent (Proposal 15) and the angle FAC is equal to angle ACG.

From here it follows that in all previous cases the sum of all three angles of spherical triangle is equal to the sum of both equal not-right angles in the quadrangle. In such a way, for each spherical triangle, where the sum of three angles S, we can find quadrangle with the same area, in which we have two right angles, two equal perpendiculars [to the base] to the sides, and each of the remaining two angles is equal to $\frac{1}{2}S$.

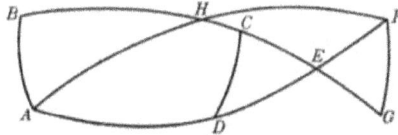
Fig. 19

Let now $ABCD$ (Fig. 19) be a spherical quadrangle, where the sides $AB = DC$ are perpendicular to BC and each of the angle at A and D is equal to $\frac{1}{2}S$. Extend the sides AD and BC till their intersection at point E and further beyond point E, make $DE = EF$ and to the extension of BC we draw a perpendicular FG. After we divide the arc BG in two equal parts and the point of division H connect by the arcs of the big circles with A and F. The triangles EFG and DCE are congruent (Proposal 15); so $FG = DC = AB$. The triangles ABH and HGF are also congruent, since they are right and have equal sides. Therefore, the arcs AH and HF belong to *the same* circle, the arc AHF is equal to π, $ADEF$ is also $= \pi$, the angle $HAD = HFE = \frac{1}{2}S - BAH = \frac{1}{2}S - HFG = \frac{1}{2}S - HFE - EFG = \frac{1}{2}S - HAD - \pi + \frac{1}{2}S$. Hence, the angle $HFE = \frac{1}{2}(S - \pi)$, or, it is also equal to the magnitude of the segment $AHFDA$, which in its turn is equal to quadrangle $ABCD$. It is easily seen, if we go from one to another, adding from the beginning to the segment the triangle EFG, and after that the triangle BAH and subtracting the equal to them triangles DCE and HFG. According to this $\frac{1}{2}(S - \pi)$ is a magnitude of the quadrangle $ABCD$, and also the magnitude of the spherical triangle, where the sum of the three angles is equal to S.

28) *If three planes intersect along parallel lines, then the sum of the three bihedral angles [formed by them] is equal to two right angles.*

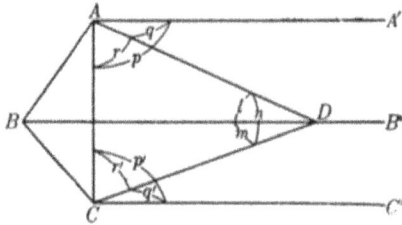

Fig. 20

Let AA', BB', CC' (Fig. 20) be parallel lines formed by intersection of the planes (Proposal 25). We take on them three arbitrary points A, B, C and imagine a plane passing through them, which therefore intersects the plane of the parallels into the straight lines AB, AC and BC. After that, through the straight line AC and some point D on the line BB' we draw one more plane, which intersects two planes parallel to AA' and BB', CC' and BB' along the lines AD and DC; the inclination of this plane to a third plane of the parallel lines AA' and CC', we will denote with w. The angles between the planes, where the parallel lines lie, we denote by X, Y, Z corresponding to the lines AA', BB', CC'; let further the right angles $BDC = a$, $ADC = b$, $ADB = c$. Imagine point A as center of spherical surface. Its intersection with the straight lines AC, AD, AA' forms a spherical triangle with sides p, q, r, whose area will be α. In it the side q is opposite the angle w, the side r – the angle X, and therefore, the side p [opposite the angle] $\pi + 2\alpha - w - X$ (Proposal 27). In such a way the straight lines CA, CD, CC' intersect the spherical surface around the center C, defined [on it] a triangle of magnitude β and with sides p', q', r' and angles: w opposite to q', Z opposite to r' and therefore, $\pi + 2\beta - w - Z$ opposite to p'. At the end, intersection of the spherical surface around D with the lines DA, DB, DC defines a spherical triangle with sides l, m, n which are opposite to angles $w + Z - 2\beta$, $w + X - 2\alpha$ and Y; hence, the magnitude of this triangle $\delta = \frac{1}{2}(X + Y + Z - \pi) - \alpha - \beta + w$. With decreasing of w the magnitude of triangles α and β is also decreasing, so $\alpha + \beta - w$ could be made smaller than any given number. In the triangle δ both sides l and m also can be reduced to zero. (Proposal 21). Hence the triangle δ can be put on one of its sides l and m on the big circle of the sphere as many times as we wish and despite that it would not fill in the semi-sphere; therefore δ will disappear together with w. From here it follows that it is necessary that $X + Y + Z = \pi$.

7 LIMITING LINE

29) *In the right triangle perpendiculars drawn from the middle of the sides, either (at all) do not meet, or all three intersect at one point.*

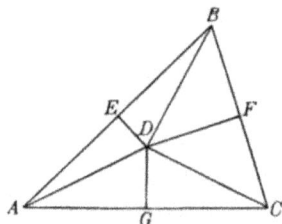

Fig. 21

Assume, that in the triangle ABC (Fig. 21) the perpendiculars ED and FD are drawn to the sides AB and BC from their middle points E and F, are intersecting in the point D; inside the angles of the triangle we draw lines DA, DB, DC.

In the equal triangles ADE and BDE (Proposal 10) $AD = BD$; we also conclude, that $BD = CD$. Therefore, ADC is isosceles triangle, and therefore the perpendicular drawn from the top D to the base AC goes in the middle point G of the latter.

The proof does not change in the case, if the point of interaction D of two perpendiculars ED and FD falls on the line AC or outside of the triangle.

If, for that reason, we accept, that two of these perpendiculars do not intersect, then the third can not meet with them.

30) *Perpendiculars drawn to the sides of rectilinear triangles from their middle points, must be mutually parallel, if we assume parallelism of only two of them.*

Let in the triangle ABC (Fig. 22) the lines DE, FG, HK, be perpendicular, drawn to its sides at their middle points D, F, H. Assume initially that the parallel perpendiculars DE and FG, which intersect the line AB at point L and M. From point L draw a straight line LG arbitrarily inside the angle BLE, which straight line should meet FG somewhere at G regardless of how small the angle of deviation LGM is (Proposal 16). As in the triangle LGM the perpendicular HK cannot meet MG (Proposal

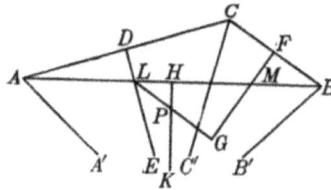

Fig. 22

29), but it should meet LG somewhere at P. From here it follows that HK should be parallel to DE (Proposal 16) and MG (Proposals 18 and 25).

If we take the sides $BC = 2a$, $AC = 2b$, $AB = 2c$ and the opposite to these sides angles denote with A, B, C then in the case just considered we have:

$$A = \Pi(b) - \Pi(c)$$

$$B = \Pi(a) - \Pi(c),$$

$$C = \Pi(a) + \Pi(b),$$

as it is easily seen that with the help of the lines AA', BB', CC', which are drawn from the points A, B, C parallel to the perpendicular HK, and hence to the two other perpendiculars DE and FG (Proposals 23 and 25).

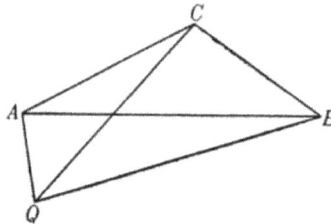

Fig. 23

Suppose now, that the parallel perpendiculars are HK and FG; in such a case a third perpendicular DE cannot intersect them (Proposal 29). Therefore, it is either parallel to them or intersect AA'. The latter assumptions means that the angle $C > \Pi(a) + \Pi(b)$. If we decrease this angle in such a manner, that it becomes equal to $\Pi(b) + \Pi(b)$ and because of this the line AC gets a new position of CQ (Fig. 23) and the magnitude of the third side BQ denote by $2c'$, then angle CBQ at the point B, which increased, should, according to what we proved above, be equal to $\Pi(a) - \Pi(c') > \Pi(a) - \Pi(c)$, from where it follows $c' > c$ (Proposal 23). However in the triangle ACQ the angles at A and Q are equal. Therefore the angle at Q in the triangle ABQ should be greater than at the point A, so $AB > BQ$ (Proposal 9); this means $c > c'$.

31) *The limiting line is called such a curved line lying in a plane, where all perpendiculars drawn from the middle point of its chords which are parallel between them.*

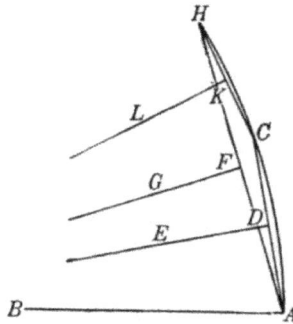

Fig. 24

In accordance with this definition we can imagine the formation of a limiting line, if we draw to a given straight line AB (Fig. 24) from a given on it point A under different angles $CAB = \Pi(a)$ the segment $AC = 2a$; the end C of such a segment will lie on the limiting line, whose points can be gradually determined. The perpendicular DE to the chord AC, drawn from its middle point D, will be parallel to the line AB, which we call *axis of the limiting line*.

In such a way each perpendicular FG, drawn from the middle point of some chord AH, will be parallel to AB. Therefore this property should belong also to each perpendicular KL, drawn from the middle of K to chord CH, between any points C and H and no limiting line can be drawn (Proposal 30). This kind of perpendiculars must be called *the axes of the limiting line* as it was done for AB.

32) *A circle, whose radius increases goes into a limiting line.*

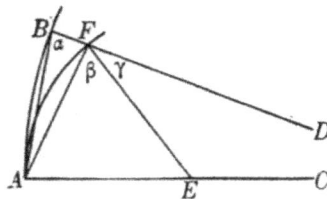

Fig. 25

Let AB (Fig. 25) be a chord of a limiting line; on its ends A and B let draw two axes AC and BD, which are forming with the chord two equal angles $BAC = ABD = \alpha$ (Proposal 31). On one of these two axis AC we take point E, call it a center of the circle and draw an arc of the circle AF from the starting point A on the axis AC till its intersection with the axis BD at point F. The corresponding point F on the radius of the circle FE forms on the one side the angle $AFE = \beta$ with a chord AF, and on the other side – the angle $FED = \gamma$ with axis BD. From here it follows, that the angle between two chords is $BAF = \alpha - \beta < \beta + \gamma - \alpha$ (Proposal 22); hence, $\alpha - \beta < \frac{1}{2}\gamma$. But the angle γ is getting smaller till zero as a result of the motion of center E when the point F does not change (Proposal 21), so as a result of point F approaching B on the axis BF if the center E does not change (Proposal 22). From here it follows that in such a decrease of the angle γ the angle $\alpha - \beta$ also disappears, i.e. the mutual inclination of two chords AB and AF, together with it the distance between the point B and the limiting line from point F on the circle also disappears.

Therefore a limiting line can be also call a *circle with a infinitely large radius*.

33) Let $AA' = BB' = x$ (Fig. 26) be two lines, parallel on the side from A to A', and their parallels are the axes of two limiting arcs (arcs of two limiting lines) $AB = s$, $A'B' = s'$; then

$$s' = se^{-x}$$

where e does not depend either from the arc s and s', or from the straight line x, i.e. from the distance of arc s' from s.

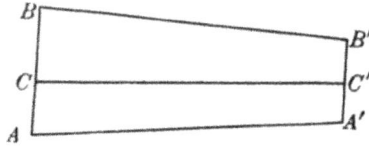

Fig. 26

In order to prove it let accept that the relation of the arc s to s' is equal to the relation of two whole numbers n and m. Between the two axes AA' and BB' draw a third axis CC', which in such a manner cuts off from the arc AB the segment $AC = t$, and from arc $A'B'$ on the same side – the segment $A'C' = t'$. Let now relation t to s is equal to the relation of two whole numbers p and q, so

$$s = \frac{n}{m}s' \qquad t = \frac{p}{q}s$$

Let us divide now s by axes in nq equal parts; then such parts will be mq on s' and np on t. Meanwhile these equal parts on s and t correspond the same equal parts on s' and t'; so,

$$\frac{t'}{t} = \frac{s'}{s}.$$

So between the two arcs, regardless where they are, t and t' between the two axes AA' and BB', the relation t to t' is the same, as long as between them we have the same distance x. If therefore for $x = 1$ take $s = es'$, then for any x it should be

$$s' = se^{-x}.$$

Since e is unknown number bound only by the requirement $e > 1$, but on the other side, a unit for measuring the length x can be chosen arbitrarily, then the latter in order to simplify the calculations can be chosen so that where e can be regarded as the basis of Neper logaritms.

Here we can also note that $s' = 0$ for $x = \infty$; hence the distance between the two parallel lines (Proposal 24) not only is getting smaller, but when extended in the side of parallelism at the end completely disappear. In such a way parallel lines have asymptotic character.

8 LIMITTING SURFACE

34) *Limiting surface* is called a surface which is formed by a rotation of a limiting line around one of its axes, which together with all other axes of the limiting lines will be also axes of the surface.

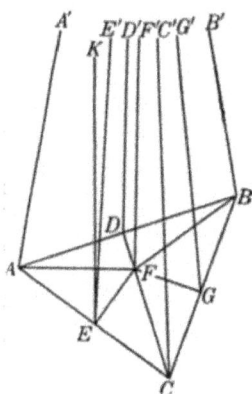

Fig. 27

Chord of [limiting surface] *inclined under equal angles to the axes, drawn through its finite points regardless of the location of these two points on the surface.*

Let A, B, C (Fig. 27) be three points on the limiting surface, let AA' be an axis of rotation, BB' and CC' – two other axes. Hence, AB and AC are chords, to whose axes inclined under equal angles are $A'AB = B'BA$, $A'AC = C'CA$ (Proposal 31); two axes BB' and CC', drawn through the end points of the third chord BC, are also parallel and lie in one plane (Proposal 25). The perpendicular DD', drawn from the middle point D of the chord AB in the plane of two parallels, should be parallel to the three axes AA', BB', CC' (Proposals 23, 25). The perpendicular EE' to the chord AC on the plane of the parallels AA', CC' will be parallel to the three axes AA', BB', CC' and to the perpendicular DD'. The angle between the planes, where the parallels AA' and BB' are located and the plane of the triangle ABC will be denoted by $\Pi(a)$, where a can be a positive, negative number or zero.

If a is positive, then we draw $FD = a$ inside the triangle ABC in its

97

plane perpendicular to the chord AB at its middle point D. If a is a negative number, then we should draw $FD = a$ outside of the triangle on the other side of the chord AB; if $a = 0$, then point F coincides with D. In all cases we get two equal right triangles AFD and DBF; hence, $FA = FB$. Now from point F we draw the perpendicular FF' to the plane of the triangle ABC.

Since the angle $D'DF = \Pi(a)$, $DF = a$, then the line FF' is parallel to DD' and to the line EE', with which it also lies in the same plane, perpendicular to the planes of the triangle ABC. Let imagine now a perpendicular EK, we draw to EF in the plane of the parallels EE' and FF', it will be also perpendicular to the plane of the triangle ABC (Proposal 13) and to the straight line AE laying in this plane (Proposal 11). Therefore the line AE is perpendicular to EK and EE', will be also perpendicular to FE (Proposal 11). The triangle AEF and FEC are equal, as a rectangle with equal sides; because $AF = FC = FB$. A perpendicular from the top F of isosceles triangle BFC to the base BC goes through the middle point of the last G. The plane, passing through this perpendicular FG and line FF', should be perpendicular to the plane of the triangle ABC and intersects the plane of the parallels BB', CC' in the line GG', which is also parallel to AA' and BB' (Proposal 25); as now CG is perpendicular to FG, and for that reason to GG', then angle $C'CG = BB'G$ (Proposal 23).

From here it follows, that for the limiting surface we can consider each of its axes as axis of the surface.

Main surface we will call each plane, drawn through the axis of the limiting surface. In accordance with this, each *main surface* cuts off the limiting surface on the limiting line, while at the other position of the cutting plane this intersection is a circle. Three main planes, intersecting each other, form with each other angles, whose sum is equal to π (Proposal 28). These angle will be viewed as angles in a limiting triangle, whose sides serve as arcs of the limiting lines, formed by the intersection of the limiting surface with these three main planes. In the limiting triangles the sides and the angles are connected through the same relations, which we have in ordinary geometry for right triangles.

9 Equations relating sides and angles of the right triangle

35) From now on we will denote the magnitude of a line by a primed letter, for example x', in order to show that the relation between x' and a line denoted by x is given by the equation below

$$\Pi(x) + \Pi(x') = \frac{1}{2}\pi$$

Let now ABC (Fig. 28) be a rectangular triangle, where the hypotenuse

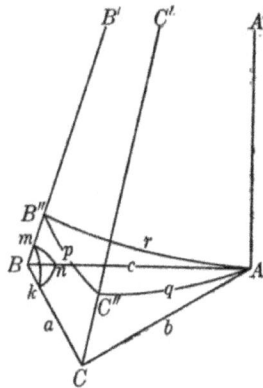

Fig. 28

$AB = c$, sides $AC = b$, $BC = a$ and their opposite angles are $BAC = \Pi(\alpha)$, $ABC = \Pi(\beta)$. At the point A we draw a perpendicular AA' to the plane of the triangle ABC and from points B and C we draw BB' and CC'' parallel to AA'. Planes, where these three parallels lie, form between each other the angle $\Pi(\alpha)$ at AA', a straight line at CC' (Proposals 11 and 13), and hence $\Pi(\alpha')$ at BB' (Proposal 28).

The intersection of the lines BA, BC, BB' with a spherical surface, described around point B as a center, defines a spherical triangle mnk, whose sides $mn = \Pi(c)$, $kn = \Pi(\beta)$, $mk = \Pi(a)$, their opposite angles are $\Pi(b), \Pi(\alpha'), \frac{1}{2}\pi$.

So together with rectangular triangle with the sides a, b, c and their opposite angles $\Pi(\alpha), \Pi(\beta), \frac{1}{2}\pi$ we assume also existence of the spherical triangle (Fig. 29) with the sides $\Pi(c), \Pi(\beta), \Pi(a)$ and the opposite angles

$$\Pi(b), \Pi(\alpha'), \frac{1}{2}\pi.$$

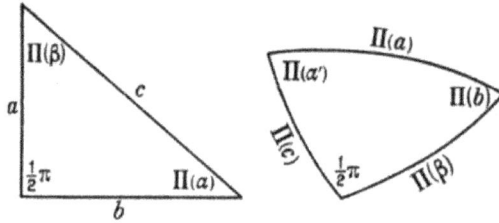

Fig. 29

Also vice-versa, for these two triangles the existence of a spherical triangle brings the existence of a rectilinear triangle, which can have sides a, α', β and their opposite angles $\Pi(b'), \Pi(c), \frac{1}{2}\pi$.

For that reason from a, b, c, α, β we can go to b, a, c, β, α and also to a, α', β, b', c.

Imagine a limiting surface, passing through point A and having for its axis the line AA' (Fig. 28). This surface intersects two other axes BB' and CC' at points B'' and C'' and its intersection with the plane of the parallels forms a limiting triangle, whose sides are $B''C'' = p, C''A = q, B''A = r$, their opposite angles are $\Pi(\alpha), \Pi(\alpha'), \frac{1}{2}\pi$, so,

$$p = r \sin \Pi(\alpha), q = r \cos \Pi(\alpha).$$

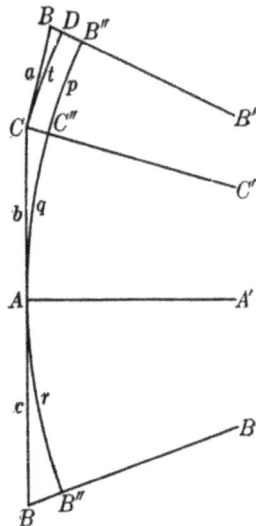

Fig. 30

Now we break the connection of the three main planes along the line BB' and arrange them in such a way in order that they together with all lines there are located in one plane. In this plane the arcs p, q, r converge in one arc of the limiting line, passing through point A and having AA' as its axis (Fig. 30). Therein on the one side of the axis AA' there will be placed the arcs q and p, the side b of the triangle, which at the point A is perpendicular to AA' – the axis CC', coming from the end of the side b parallel to AA' and passing through the point of connection C'' of the arcs p and q. The side a is perpendicular to CC' at the point C, and also coming from the end of that side of the axis BB', and is parallel to AA' and passing through the end B'' of the arc p. On the other side of AA' there will be lying: the side c, perpendicular to AA' at the point A, and the axis BB', parallel AA' and coming from the end of the side b through the end point B'' of the arc r. The magnitude of the side CC'' depends on b, whose dependence we will express through $CC'' = f(b)$. In such away will be $BB'' = f(c)$.

If we regard CC'' as an axis, we draw from point C a new limiting line until the intersection D with the axis BB' and denote the arc CD by t, then

$$BD = f(a), \quad BB'' = BD + DB'' = BD + CC''$$

and therefore,

$$f(c) = f(a) + f(b).$$

Except this, we notice (Proposal 33), that

$$t = pe^{f(b)} = r \sin \Pi(\alpha) e^{f(b)}.$$

If the perpendicular to the plane of the triangle ABC (Fig. 28) had been drawn not at point A, but at point B, then the lines c and r would have remain the same, the arcs q and t would have gone in t and q, the straight lines a and b in b and a, and the angle $\Pi(\alpha)$ would have been replaced by $\Pi(\beta)$; therefore, we should have had

$$q = r \sin \Pi(\beta) e^{f(a)}$$

substituting instead of q its value, we find

$$\cos \Pi(\alpha) = \sin \Pi(\beta) e^{f(a)},$$

further, replacing α and β with b' and c:

$$\sin \Pi(b) = \sin \Pi(c) e^{(a)},$$

and multiplying by $e^{f(b)}$:

$$\sin \Pi(b) e^{f(b)} = \sin \Pi(c) e^{f(c)}$$

From here it follows as well:

$$\sin \Pi(a) e^{f}(a) = \sin \Pi(b) e^{f(b)}.$$

As the straight lines a and b do not depend on each other, but on the other side, when $b = 0$, $f(b) = 0$, $\Pi(b) = \frac{1}{2}\pi$, then for any straight line a

$$e^{-f(a)} = \sin \Pi(a);$$

according to this

$$\sin \Pi(c) = \sin \Pi(a) \sin \Pi(b)$$

$$\sin \Pi(\beta) = \cos \Pi(\alpha) \sin \Pi(a).$$

As a result of changing the letters we get

$$\sin \Pi(\alpha) = \cos \Pi(\beta) \sin \Pi(b),$$
$$\cos \Pi(b) = \cos \Pi(c) \cos \Pi(\alpha),$$
$$\cos \Pi(a) = \cos \Pi(c) \cos \Pi(\beta).$$

If in the spherical rectilinear triangle (Fig. 29) we denote the sides $\Pi(c), \Pi(\beta)$, $\Pi(a)$ with their opposite angles $\Pi(b), \Pi(\alpha')$ by the letters a, b, c, A, B then the newly found equations have the form, as we know, and as it is known, for right triangles in spherical trigonometry, namely:

$$\sin a = \sin c \sin A,$$
$$\sin b = \sin c \sin B,$$
$$\cos A = \cos a \sin B,$$
$$\cos B = \cos b \sin A,$$
$$\cos c = \cos a \cos b;$$

from these equations we can go to the equations of any spherical triangles. In such a way, spherical trigonometry does not depend on whether the sum of the three angles in a rectilinear triangle is equal to two right angles or not.

10 FINDING THE FUNCTION $\Pi(x)$

36) Let us now consider again the rectilinear right triangle ABC (Fig. 31), whose sides are a, b, c and the opposite angles $-$ $\Pi(\alpha), \Pi(\beta), \frac{1}{2}\pi$. Extend the hypotenuse c beyond point B and make $BD = \beta$; from point D we draw to BD a perpendicular DD', which therefore will be parallel to BB', i.e., an extension of the side a beyond point B. From point A we draw to DD' a parallel AA', which at the same time will be parallel to CB' (Proposal 25). Therefore angle $A'AD = \Pi(c + \beta)$, $A'AC = \Pi(b)$; so,

$$\Pi(b) = \Pi(\alpha) + \Pi(c + \beta)$$

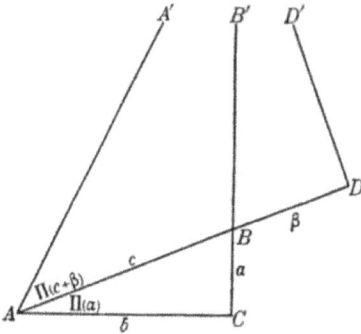

Fig. 31

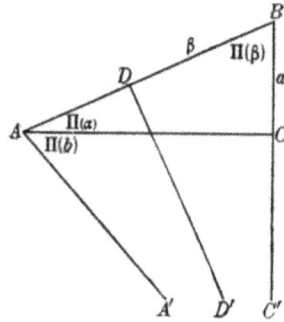

Fig. 32

If we draw β on the hypotenuse c from point B, after that from the end point D (Fig. 32) we draw to AB inside the triangle a perpendicular DD' and from point A we draw AA' parallel to DD', then BC with its extension CC' will be a third parallel; then the angle $CAA' = \Pi(b)$, $DAA' = \Pi(c - \beta)$, so,

$$\Pi(c - \beta) = \Pi(\alpha) + \Pi(b).$$

The last equation remains valid in the case, when $c = \beta$ or $c < \beta$. If $c = \beta$ (Fig. 33), then the perpendicular AA', drawn to AB from point A, is parallel to the side $BC = a$ with its extension CC'. Hence $\Pi(\alpha) + \Pi(b) = \frac{1}{2}\pi$ at the same time $\Pi(c - \beta) = \frac{1}{2}\pi$ (Proposal 23). If $c < \beta$ (Fig. 34), then

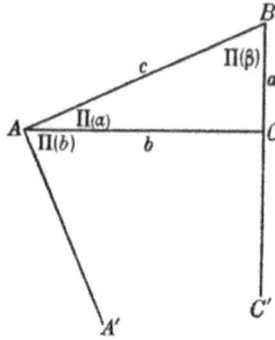

Fig. 33

the end of the segment β goes on the other side of point A at D on the extension of the hypotenuse AB. The drawn from here perpendicular DD' to AD and the parallel line AA' to it from point A will be also parallel to the side $BC = a$ with its extension CC'. Here the angle $DAA' = \Pi(\beta - c)$, therefore, $\Pi(\alpha) + \Pi(b) = \pi - \Pi(\beta - c) = \Pi(c - \beta)$ (Proposal. 23).

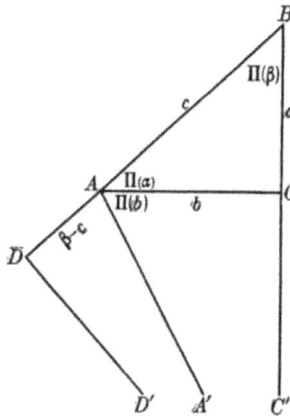

Fig. 34

Connecting the two found equations, we get

$$2\Pi(b) = \Pi(c - \beta) + \Pi(c + \beta),$$

$$2\Pi(\alpha) = \Pi(c - \beta) - \Pi(c + \beta),$$

from here it follows

$$\frac{\cos \Pi(b)}{\cos \Pi(\alpha)} = \frac{\cos[\frac{1}{2}\Pi(c - \beta) + \frac{1}{2}\Pi(c + \beta)]}{\cos[\frac{1}{2}\Pi(c - \beta) - \frac{1}{2}\Pi(c + \beta)]}.$$

If here we put the value (Proposal 35)

$$\frac{\cos \Pi(b)}{\cos \Pi(\alpha)} = \cos \Pi(c),$$

we get

$$\tan \frac{1}{2}\Pi(c)^2 = \tan \frac{1}{2}\Pi(c - \beta) \tan \frac{1}{2}\Pi(c + \beta).$$

Here β is an arbitrary number, because the angle $\Pi(\beta)$ with one side [hypotenuse] c can be chosen arbitrarily in the range from 0 till $\frac{1}{2}\pi$; so β [it can be chosen] arbitrarily between 0 and ∞; taking here in turn of the angles and the sides $\beta = c, 2c, 3c$ and so on, for each positive number n we will get:

$$\left[\tan \frac{1}{2}\Pi(c)\right]^n = \tan \frac{1}{2}\Pi(nc).$$

If we consider here n as a relation of two lines x and c and accept, that,

$$c \tan \frac{1}{2}\Pi(c) = e^c,$$

then we will get in general for any line x, for both positive and negative,

$$\tan \frac{1}{2}\Pi(x) = e^{-x}, \tag{Π}$$

where e could be any number, greater then unity, because $\Pi(x) = 0$ at $x = \infty$.

As far as, the unit, with which we measure lines, is arbitrary, then for e can be also accepted the base of natural (Neper) logarithms.

11 EQUATIONS RELATING THE SIDES AND ANGLES OF ANY TRIANGLE

37) From the equations found above it is sufficient to know the following two equations:

$$\sin \Pi(c) = \sin \Pi(a) \sin \Pi(b),$$

$$\sin \Pi(\alpha) = \sin \Pi(b) \cos \Pi(\beta),$$

relating the last to both sides a and b, in order that from their connection we draw two remaining (Proposal 35) and without ambiguity concerning the algebraic signs, or angles here are astute. In the same way, we arrive to two equations:

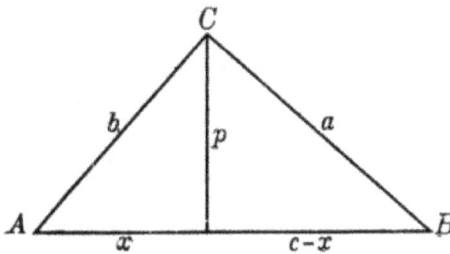

Fig. 35

1. $\tan \Pi(c) = \sin \Pi(\alpha) \tan \Pi(a),$

2. $\cos \Pi(a) = \cos \Pi(c) \cos \Pi(\beta).$

Consider now rectilinear triangle with sides a, b, c, and opposite angles A, B, C (Fig. 35). If A and B are astute angle, then the perpendicular p from the top of angle C goes inside the triangle and divides the side c in two parts, namely: a part x on the side of the angle A and $c - x$ on the side of the angle B. In such a way, we get two right triangles, applying to them equation (1), we get

$$\tan \Pi(a) = \sin B \tan \Pi(p),$$

$$\tan \Pi(b) = \sin A \tan \Pi(p).$$

which equations remain without changes, even if one of the angles, for example B, had been right (Fig. 36) or obtuse (Fig. 37). In such a way, we get a common relation for all triangles

$$3. \ \sin A \tan \Pi(a) = \sin A \tan \Pi(a) = \sin B \tan \Pi(b).$$

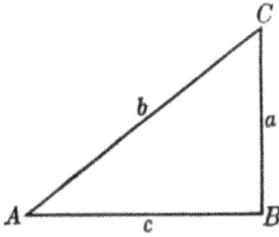

Fig. 36

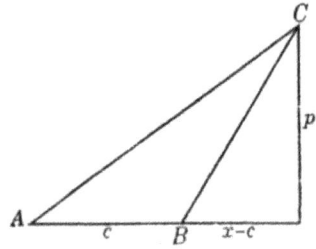

Fig. 37

For the triangle with astute angles A and B (Fig. 35) we get also [equation (2)]

$$\cos \Pi(x) = \cos A \cos \Pi(b),$$
$$\cos \Pi(c - x) = \cos B \cos \Pi(a),$$

[a]

which equations are valid also for triangles, where angles A or B are right or obtuse. For example, when $B = \frac{1}{2}\pi$ (Fig. 36) it is necessary to have $x = c$; the first equation goes into the equation, which we obtained above [equation (2)], and the second equation is self-satisfied. When we have $B > \frac{1}{2}\pi$ (Fig. 37) the first equation remains without changes, instead of the second equation we should write the relation

$$\cos \Pi(x - c) = \cos \pi - B) \cos(\Pi(a);$$

but $\cos \Pi(x - c) = -\cos \Pi(c - x)$ (Proposal 23), and also

$$\cos(\pi - B) = -\cos B.$$

If A is right or obtuse angle, then instead of x and $c - x$ it is necessary to take $c - x$ and x, and this case will be transformed into the previous.

In order that from these two equations we exclude x, we notice, that (Proposal 36)

$$\cos \Pi(c - x) = \frac{1 - \tan \frac{1}{2}\Pi(c - x)^2}{1 + \tan \frac{1}{2}\Pi(c - x)^2} = \frac{1 - e^{2x - 2c}}{1 + e^{2x - 2c}}$$

[b]

$$= \frac{1 - \tan \frac{1}{2}\Pi(c)^2 \cot \frac{1}{2}\Pi(x)^2}{1 + \tan \frac{1}{2}\Pi(c)^2 \cot \frac{1}{2}\Pi(x)^2} = \frac{\cos \Pi(c) - \cos \Pi(x)}{1 - \cos \Pi(c) \cos \Pi(x)}.$$

If we substitute here the expressions for $\cos \Pi(x)$, $\cos \Pi(c-x)$, then we get

$$\cos \Pi(c) = \frac{\cos \Pi(a) \cos B + \cos \Pi(b) \cos A}{1 + \cos \Pi(a) \cos \Pi(b) \cos A \cos B}, \qquad [c]$$

from where it follows

$$\cos \Pi(a) \cos B = \frac{\cos \Pi(c) - \cos A \cos \Pi(b)}{1 - \cos A \cos \Pi(b) \cos \Pi(c)}$$

and at the end,

$$\sin \Pi(c)^2 = [1 - \cos B \cos \Pi(c) \cos \Pi(a)] \cdot [1 - \cos A \cos \Pi(b) \cos \Pi(c)].$$

In such a way, it should have been

4. $\sin \Pi(a)^2 = [1 - \cos C \cos \Pi(a) \cos \Pi(b)] \cdot [1 - \cos B \cos \Pi(c) \cos \Pi(a)]$,

   $\sin \Pi(b)^2 = [1 - \cos A \cos \Pi(b) \cos \Pi(c)] \cdot [1 - \cos C \cos \Pi(a) \cos \Pi(b)]$.

From these three equations we also receive

$$\frac{\sin \Pi(b)^2 \sin \Pi(c)^2}{\sin \Pi(a)^2} = [1 - \cos A \cos \Pi(b) \cos \Pi(c)]^2.$$

From here without ambiguity with respect to the sign we have:

5. $\cos A \cos \Pi(b) \cos \Pi(c) + \dfrac{\sin \Pi(b) \sin \Pi(c)}{\sin \Pi(a)} = 1.$

If here with agreement with equation (3) we substitute the value of $\sin \Pi(c)$:

$$\sin \Pi(c) = \frac{\sin A}{\sin C} \tan \Pi(a) \cos \Pi(c),$$

we get

$$\cos \Pi(c) = \frac{\cos \Pi(a) \sin C}{\sin A \sin \Pi(b) + \cos A \sin C \cos \Pi(a) \cos \Pi(b)}$$

or we substitute this expression for $\cos \Pi(c)$ in equation (4),

6. $\cot A \sin C \sin \Pi(b) + \cos C = \dfrac{\cos \Pi(b)}{\cos \Pi(a)}.$

Excluding from here $\sin \Pi(b)$ with the help of equation (3), we get

$$\frac{\cos \Pi(a)}{\cos \Pi(b)} \cos C = 1 - \frac{\cos A}{\sin B} \sin C \sin \Pi(a).$$

In the mean time, changing letters in equation (6) gives

$$\frac{\cos \Pi(a)}{\cos \Pi(b)} = \cot B \sin C \sin \Pi(a) + \cos C.$$

From two of the last equations it follows

$$7. \quad \cos A + \cos B \cos C = \frac{\sin B \sin C}{\sin \Pi(a)}.$$

All four equations, showing the dependence between sides a, b, c and their opposite angles A, B, C, according to this (equations (3), (5), (6), (7)) will be

$$8. \quad \left\{ \begin{array}{l} \sin A \tan \Pi(a) = \sin B \tan \Pi(b), \\ \cos A \cos \Pi(b) \cos \Pi(c) + \frac{\sin \Pi(b) \sin \Pi(c)}{\sin \Pi(a)} = 1, \\ \cot A \sin C \sin \Pi(b) + \cos C = \frac{\cos \Pi(b)}{\cos \Pi(a)}, \\ \cos A + \cos B \cos C = \frac{\sin B \sin C}{\sin \Pi(a)}. \end{array} \right.$$

If the sides of the triangle a, b, c are very small, then we can use approximate values (Proposal 36):

$$\cot \Pi(a) = a,$$
$$\sin \Pi(a) = 1 - \frac{1}{2} a^2,$$
$$\cos \Pi(a) = a$$

and analogically for the other sides b and c. The equations (8) for such triangles transform in the following:

$$b \sin A = a \sin B,$$
$$a^2 = b^2 + c^2 - 2bc \cos A,$$
$$a \sin(A + C) = b \sin A,$$
$$\cos A + \cos(B + C) = 0.$$

From these equations the two first equations are accepted in the ordinary geometry; the last two, with the help of the two first, bring the conclusion that

$$A + B + C = \pi.$$

In such a way, the imaginary geometry transforms into the ordinary, if we assume that the sides of rectilinear triangle are very small.

I published some studies about measurements of curved lines, areas of plane figures, surfaces and volumes of bodies, also about applying the imaginary geometry to calculus in "The learned notes of the University of Kazan."

Equations (8) already themselves are a necessary base for considering the hypotheses of imaginary geometry as possible. So we do not have any other means except astronomical observations to judge about accuracy, which gives calculations of ordinary geometry.

As I showed in one of my works this accuracy extends far so, for example, in the triangle, whose sides are amenable to our measurements, the sum of the angles does not deviate from two right angles even by a hundredth part of a second.

It is wonderful also, that equations (8) of the plane geometry transforms into the equations of spherical triangles, if instead of the sides a, b, c substitute $a\sqrt{-1}$, $b\sqrt{-1}$, $c\sqrt{-1}$; and in such a change, however, it is necessary also to write the following equations

$$\sin \Pi(a) = \frac{1}{\cos a},$$

$$\cos \Pi(a) = \sqrt{-1}\tan a,$$

$$\tan \Pi(a) = \frac{1}{\sin a\sqrt{-1}}$$

similar changes we should make for the sides b, c; in this way we will go from equation (8) to the following;

$$\sin A \sin b = \sin B \sin a,$$

$$\cos a = \cos b \cos c + \sin b \sin c \cos A,$$

$$\cot A \sin C + \cos C \cos b = \sin b \cot a,$$

$$\cos A = \cos a \sin B \sin C - \cos B \cos C.$$

PANGEOMETRY
1855

1 INTRODUCTION

1. We do not have enough concepts in geometry on which to draw the proof of the theorems. The sum of the three angles of rectilinear triangles is equal to two right angles. A theorem in whose truthfulness nobody had any doubts up to now, because it had not been any controversies in the conclusions, which are drawn from here, and for that reason the measurement of the angles in rectilinear triangles are in agreement within the limits of the errors in the most precise measurements with this theorem. The lack of postulates for proving the given theorems have forced geometricians to assume explicitly or implicitly auxiliary provisions, which regardless of their simplicity are at the same time arbitrary and so they cannot be assumed. For example, we accept: that a circle with extremely great radius goes into right line, and a sphere with infinitely large radius – into a plane; that the angles of a rectilinear triangle depend only on the content of the sides, but not on the sides themselves, or at the end as this is assumed in the foundations of geometry, that from a given point on a plane we cannot draw more than one straight line parallel to a given straight line in this plane, when all the other straight lines, drawn from the same point and in the same plane, should necessary on a given extension intersect the given straight line. Under a line parallel to another, we understand a straight line whose extension in both sides will never meet the line with which it is a parallel. This definition by itself is not sufficient, because it does not point to a single line. The same can be said about most of the content of the definition, given in the foundations of geometry, because these definitions not only did not indicate the origin of geometrical quantity, which should be defined, but even did not prove that such quantities can exist. In such a way, we define some properties of the straight line and the plane; it is known that straight lines are these lines which merge if they have two common points; that the plane is such kind of a surface, where the straight line entirely lies, drawn through two points on the plane.

Instead of beginning geometry of a straight line and a plane, as usual I prefer to begin with a sphere and a circle, whose definitions are not subjected to incompleteness, because these definitions contain the origin of the quantities. After that I define the plane as a surface where equal spheres intersect, described around two fixed points. At the end I define a straight line as an intersection of equal circles in the plane, described around two two fixed points on the same plane. Having assumed such definitions the whole theory of straight lines and perpendicular planes can be explained strictly and very easily.

A straight line, drawn from a given point on the plane, I call *parallel* to a given straight line in the same plane. It will form a boundary between these straight lines, drawn from the same point on the same plane, which intersect the given straight line on sufficient extension and these which do not intersect, regardless of their extension. This side, where we have intersection I call *side of parallelism*.

I published a complete theory of parallel under the title "Geometrische zur Theorie der Parallelinien Berlin 1840. In der Finke'schen Buchhandlung ". In this work I gave proofs of all suggestions where it is not necessary to use parallel lines. It is very important to acknowledge (Geometr. Untersuchungen 27) between these suggestions proposals this, which give relation between the surface of spherical triangle to the whole sphere. If A, B, C mean angles of the spherical triangle, then the content of the surface of this spherical triangle to the surface of the whole sphere, to which it belongs, will be equal to the equation

$$\frac{1}{2}(A + B + C - \pi)$$

to four right angles. Here π means two right angles. After I prove, that the sum of the three angles in the rectilinear triangle can not be more than two right angles. (Geometr. Unters. §19), and if this sum is equal to two right angles in some rectilinear triangle, then it should be the same in all rectilinear triangles. (Geometr. Unters. §20).

So there are only two possible assumptions: either the sum of three angles in each rectilinear triangle is equal to two right angles – this proposal forms the ordinary geometry – or in each rectilinear triangle this sum is less then two right angles, and the last suggestion is the base of the special geometry, which I called imaginary geometry, which could be also call *Pangeometry*, because this notion means geometry of a wider kind, where the ordinary geometry will be a special case.

2 FUNDAMENTAL ASSUMPTIONS

2. From the accepted foundations of pangeometry it follows, that a perpendicular p, drawn from a point on the straight line to the parallel to that line, forms with the first line two angles, one of which is astute. I call this angle *angle of parallelism*, and the side of the first of these straight lines, where the astute angle is, and which remains the same for all points on the straight line is the *side of parallelism*. I will denote this angle $\Pi(p)$ (Fig. 1), because it depends on the length of the perpendicular p. In the ordinary geometry $\Pi(p) = \frac{\pi}{2}$ for every p. In pangeometry the angle $\Pi(p)$ can take on all numbers from 0, which correspond $p = \infty$, to $\frac{\pi}{2}$ which correspond $p = 0$. (Geometr. Unters. §23). In order that we give the function $\Pi(p)$ analytical value more generally, I accept that, for a negative p, where the first definition is not applicable, the value of this function is given by the equation: $\Pi(p) + \Pi(-p) = \pi$.

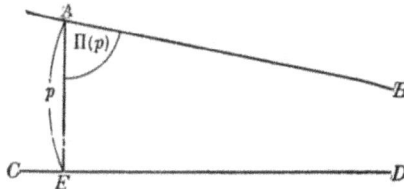

Fig. 1

So for all angles $A > 0$ and $< \pi$ it can be found a line p such, that $\Pi(p) = A$, where the line p is positive, if $A < \frac{\pi}{2}$. Conversely, for any line p there exists an angle A so, that $A = \Pi(p)$.

I call a *limiting circle* such a circle whose radius is infinitely large; it can be drawn gradually by points belonging to it. We take point $[M]$ on the given straight line (Fig. 2), and call this point *top* and the straight line itself $[MM']$ – *axis* of the limiting circle; after that we construct on this straight line an angle $A \geq 0$ and $\leq \frac{\pi}{2}$ whose top coincide with the top of the limiting circle, and one of its sides coincides with the axis. At the end, a – a line, which gives $\Pi(a) = A$ and place it on the other side of the angle A from the top of the straight line $2a$, the end $[N]$ this straight line will lie on the limiting circle. In order to extend the limiting circle on the other side of the axis, we must repeat the same construction on this side of the axis. From here, it is seen that all straight lines, which are parallel to the axis of limiting circle, can be equally treated as axes of the limiting circle.

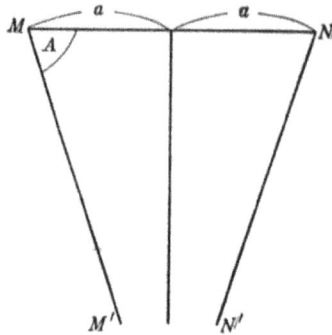

Fig. 2

A rotation of the limiting circle around one of its axes forms a surface, which I call a *limiting sphere*, a surface which therefore will be a limit of approximation for the sphere with increasing of its radius to infinity.

We will call the axis of rotation, and therefore all lines parallel to the axis of rotation, *axis* of the limiting sphere, and *radial plane* will call a plane, in which one or several axes of rotation lie. The intersection of the limiting sphere with the radial plane gives a limiting circle. Part of the limiting sphere, bounded by three arcs of the limiting circle, is called a *limiting spherical triangle*. The arcs of the limiting circle will be called sides of the limiting spherical triangle, and the plane angles between the planes, where the arcs of the limiting circle lie, will be called angles of the limiting spherical triangle.

Two straight lines, parallel to a third straight line, are also parallel to each other (Geometr. Unters. §25). From here it follows, that all axes of the limiting circle and limiting sphere are also parallel to each other.

If three planes intersect by two in three parallel straight lines, and if each plane is bounded between two parallels, then the sum of the three plane angles, which are formed by these planes, is equal to two right angles (Geometr. §28).

From this assumption it follows, that the sum of three angles in the limiting spherical triangle is equal to two right angles, and all this, that what in the ordinary geometry proves the content of the sides of the rectilinear triangle, can be repeated and proved in pangeometry for the sides of the limiting spherical triangle. It should only replaced parallel straight lines with one of the sides of the triangle by arcs of the limiting circle, drawn through the point on one of the sides of the limiting spherical triangle under the same angle with this side. For example, if p, q, r are sides of the limiting spherical right triangle and $P, Q, \frac{\pi}{2}$ – angles opposite these sides, then we should take as well as for rectilinear rectangular triangle in the ordinary geometry the following equations;

$$p = r \sin P = r \cos Q,$$
$$q = r \cos P = r \sin Q,$$
$$P + Q = \frac{\pi}{2}.$$

3. It is proven in the ordinary geometry, that the distance between two parallel straight lines is everywhere equal; it is the opposite in pangeometry, the distance of p between points on one straight line to a parallel line with gets smaller on the side of parallelism, i.e. on the side facing the angle of parallelism $\Pi(p)$.

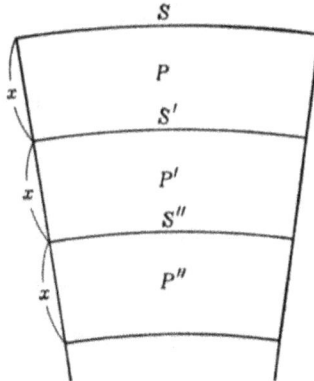

Fig. 3

Now let S, S', S'' and so on – a sequence of arcs of the limiting circles between between two parallel straight lines, which serve as axes of these limiting circles (Fig. 3). Assume that the parts of parallel lines between the arcs are consecutively equal and are measured by the mutual distances of two consecutive arcs x. We call E the content of S to S', when x is equal to a unit of length

$$\frac{S}{S'} = E,$$

where E is a positive number and greater then one. Let now number E be expressed through the content of two whole numbers n to m, so $E = \frac{n}{m}$;we divide the arc S into m equal parts; through the dividing points introduce parallel straight lines to the axes of the limiting circles. These parallels divide arcs (each of them) $S', S'',\ldots$ each into m equal parts. We transfer the area between the arcs S', S'' on the area between the arcs S, S' and put the arc S' on the S, and therefore the arc S'' on S', repeating the arc $\frac{S'}{m}$; it should be superimposed n times on the arc S.

Parallelism of the straight lines forces the arc $\frac{S''}{m}$ to be superimposed on the arc S' also n times; hence,

$$\frac{S}{S'} = \frac{S'}{S''}.$$

In order to prove the above in the case of incommensurability of the number E, we can use the same methods, which are used usually in similar cases in geometry; I will omit these details for the sake of shortness. So,

$$\frac{S}{S'} = \frac{S'}{S''} = \frac{S''}{S'''} = \ldots = E.$$

It is necessary that for each line x

$$S' = SE^{-x},$$

where E is a number, equal to the content of S to S', when $x = 1$.

It is necessary to notice, that the content E does not depend on the length of the arc S and does not change, if two given parallel straight lines converge and move away from each other remaining still parallel. The number E, which should be greater than unity, depends only on that straight line, which is chosen for unit for the measurement of straight lines and which measures the distance between two subsequent arcs and can be taken arbitrarily.

The property, which we proved concerning the arcs $S, S', S'', \ldots$ remains for the areas $P, P', P'' \ldots$, as well, bounded by consecutive arcs and parallel straight lines. So,

$$P' = PE^{-x}.$$

If we connect n similar areas in a series, then the sum will be

$$P\frac{1 - E^{-nx}}{1 - E^{-x}}.$$

For $n = \infty$ this expression gives the area between two parallel lines, bounded on the one side by the arc S and unbounded on the side of parallelism; the value of area will be the following:

$$\frac{P}{1 - E^{-x}}.$$

If we choose for a unit area this area, which corresponds to the arc $S = 1$ and distance $x = 1$, then the found expression for the area applies to each arc S

$$= \frac{ES}{E - 1}.$$

In the ordinary geometry number E is constant and equal to unity; in the ordinary geometry two straight parallel lines are at equal distances from each other everywhere, and the area between the two parallel lines, bounded only on one side by a common perpendicular to it, is infinitely large.

3 EQUATIONS RELATING SIDES AND ANGLES OF THE RIGHT TRIANGLE; FORMULAS OF SPHERICAL TRIGONOMETRY

4. Let us consider now a rectilinear and right triangle, with sides a, b, c and angles $A, B, \frac{\pi}{2}$, which are opposite to the sides. The astute angles A, B can be taken for angles of parallelism $\Pi(\alpha), \Pi(\beta)$, and the corresponding two straight lines α, β are positive. Let us denote by an accent over a letter such a straight line for which the angle of parallelism serves as a complementary to the right angle of the angle of parallelism, which corresponds to the straight line, which is denoted by the same letter but without an accent. In such a way,

$$\Pi(\alpha) + \Pi(\alpha') = \frac{\pi}{2}, \quad \Pi(\beta) + \Pi(\beta'). = \frac{\pi}{2}$$

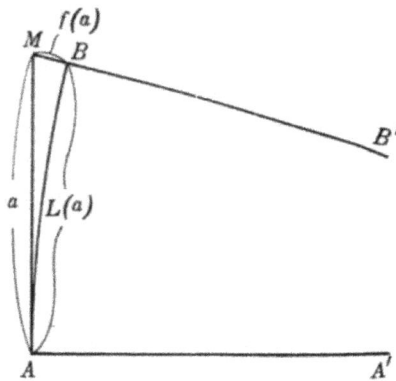

Fig. 4

Imagine a perpendicular a (Fig. 4) to the axis of the limiting circle at the top of the axis. Through the top of perpendicular a draw a straight line, parallel to the axis on the side of parallelism. Denote by $f(a)$ a part of this parallel line between the top of the perpendicular and the arc, we denote by $L(a)$ the length of the arc of the limiting circle, cut off by a parallel line to the top of the limiting circle. In the ordinary geometry $L(a) = a$; $f(a) = 0$ for each lines a.

We draw perpendicular AA' to the plane of a rectilinear rectangular triangle, whose sides we called a, b, c and let the perpendicular AA' passes through the top of the angle A. Draw through this perpendicular two planes: one, which we call first, through side b, and another which we call second plane, through side c (Fig. 5).

Draw in the second plane through the top B the angle $\Pi(\beta)$ of the straight line BB', parallel to AA'. The third plane we draw through BB' and through the side a of the triangle.

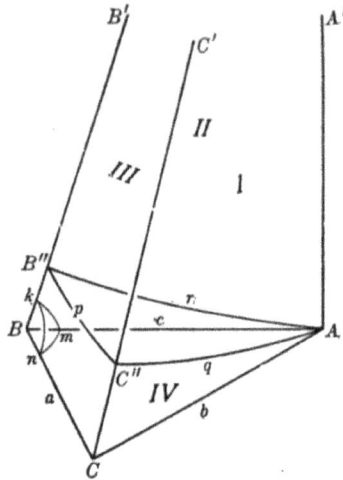

Fig. 5

This third plane intersects the first plane in the straight line CC', parallel to AA'. Imagine now a sphere, described around point B as center of an arbitrary radius smaller than a. This sphere intersects, hence, the side a with the triangle and the straight line BB' in three points, which we will denote: first n, second m, third k. The arcs of the greater circle, which are formed from the intersection of the sphere with the three planes, passing through B, which connect the points n, m, k in pairs, will form a spherical triangle with a right angle at m and whose sides will be $mn = \Pi(\beta)$; $km = \Pi(c)$; $kn = \Pi(a)$. The angle knm of the spherical triangle is equal to $= \Pi(b)$, and the angle $kmn = \frac{\pi}{2}$. Three straight lines AA', BB', CC' which are parallel among themselves, form a sum of three plane angles $= \pi$. From here it follows, that the third angle mkn of the spherical triangle $= \Pi(\alpha')$.

So, to each rectilinear rectangular triangle, with sides a, b, c with opposite angles $\Pi(\alpha), \Pi(\beta), \frac{\pi}{2}$ there corresponds a spherical rectangular triangle whose sides $\Pi(\beta), \Pi(c), \Pi(a)$ with the opposite angles $\Pi(\alpha')$, $\Pi(b)$, $\pi/2$ (Fig. 6).

Construct a rectilinear rectangular triangle, whose perpendicular sides

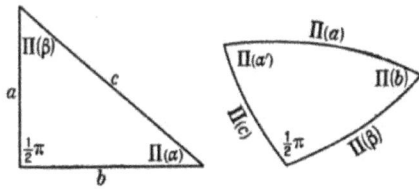

Fig. 6

will be α', a, hypotenuse $- g$, and $\Pi(\lambda)$ is the angle opposite to the side a and $\Pi(\mu)$ – the angle opposite to the side α'.

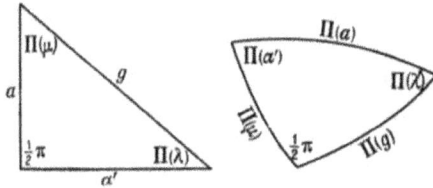

Fig. 7

We go from this triangle to a spherical, in a similar way as we did from rectilinear triangle ABC to the spherical triangle kmn. The sides of the new spherical triangle (Fig. 7) will be

$$\Pi(\mu), \Pi(g), \Pi(a)$$

with opposite angles

$$\Pi(\lambda'), \Pi(\alpha'), \frac{\pi}{2}.$$

There will be parts in it respectively equal to the parts of the spherical triangle kmn, because the sides of the latter triangle are $\Pi(c), \Pi(\beta), \Pi(a)$ with opposite angles

$$\Pi(b), \Pi(\alpha'), \frac{\pi}{2}.$$

This proves that these two spherical rectilinear triangles have equal hypotenuses and one of the adjacent angles is equal, and hence they are equal. From here it follows

$$\mu = c, \quad g = \beta, \quad b = \lambda'.$$

So the existence of a rectilinear rectangular triangle with sides

$$a, b, c$$

and with opposite angles

$$\Pi(\alpha), \Pi(\beta), \frac{\pi}{2}$$

implies the existence of a rectilinear rectangular triangle, whose sides a, α', β with opposite angles:

$$\Pi(b'), \Pi(c), \frac{\pi}{2}.$$

The same can be expressed, assuming that if

$$a, b, c, \alpha, \beta$$

are parts of a rectilinear rectangular triangle then a, α', β, b', c will be corresponding parts of another rectilinear right triangle.

5. Imagine a limiting sphere through point A of the given rectilinear rectangular triangle, in which sphere the perpendicular AA' to the plane of this triangle serves as an axis, and point A as a top (Fig. 5) then we get a limiting spherical triangle from the intersection of the limiting sphere and the three planes, drawn through the straight lines AA', BB', CC'.

Let denote the side of this limiting spherical triangle p, q, r in such a way, that p be the intersection of the limiting sphere $[c]$ with the plane, which passes through a; q – the intersection of the limiting sphere with the plane which passes through b, and r – the intersection of the sphere with the plane, which passes through c; the angles opposite to these sides will be: $\Pi(\alpha)$ opposite to p, $\Pi(\alpha')$ opposite to q, $\frac{\pi}{2}$ opposite to r. According to the above notations, here $q = L(b); r = L(c)$. The limiting sphere intersects the straight line CC' at the point, whose distance from C will be, according to the same notations, $f(b)$; in a similar way $f(c)$ will be the distance from the point of intersection of the limiting sphere with the straight line BB' from point B.

It is easy to see that

$$f(b) + f(a) = f(c).$$

In the triangle whose sides are the limiting arcs p, q, r, will be

$$p = r \sin \Pi(\alpha); \quad q = r \cos \Pi(\alpha).$$

Multiplying the first from these two equations by $E^{(b)}$, we get:

$$pE^{f(b)} = r \sin \Pi(\alpha) E^{f(b)};$$

but

$$pE^{F(b)} = L(a),$$

and hence

$$L(a) = r \sin \Pi(\alpha) E^{f(b)}.$$

In this way, we will get

$$L(b) = r \sin \Pi(\beta) E^{f(a)}.$$

Together with this $q = r \cos \Pi(\alpha)$ or, which is equal :

$$L(b) = r \cos \Pi(\alpha).$$

Comparing these two notations $L(b)$ gives the equation

$$\cos \Pi(\alpha) = \sin \Pi(\beta) E^{f(a)}; \tag{1}$$

replacing here α with b', β with c and leaving a without a change, as this will allow us to get as it was proved above

$$\cos \Pi(b') = \sin \Pi(c)\, E^{f(a)};$$

or as far as

$$\Pi(b) + \Pi(b') = \frac{\pi}{2},$$

$$\sin \Pi(b) = \sin \Pi(c) E^{f(a)}.$$

In such a way we will have

$$\sin \Pi(a) = \sin \Pi(c)\, E^{f(b)}.$$

Multiplying the last equation by $E^{f(a)}$ and putting here $f(c)$ instead of $f(a) + f(b)$; so we find

$$\sin \Pi(a)\, E^{f(a)} = \sin \Pi(c)\, E^{f(c)}.$$

But as the perpendicular sides in the rectilinear rectangular triangle can be changed when the hypotenuse remains the same, then in this equation we can set $a = 0$ without changing c; this gives $f(0) = 0$ and $\Pi(0) = \frac{\pi}{2}$,

$$1 = \sin \Pi(c)\, E^{f(c)}$$

or

$$E^{f(c)} = \frac{1}{\sin \Pi(c)} \quad \text{for each line of } c.$$

Now we take equation (1)

$$\cos \Pi(\alpha) = \sin \Pi(\beta)\, E^{f(a)}$$

and put here $\frac{1}{\sin \Pi(a)}$ at the place of $E^{f(a)}$; then we will get the following equation:

$$\cos \Pi(\alpha) \sin \Pi(a) = \sin \Pi(\beta) \tag{2}$$

Changing here α, β to b', c, leaving a without a change, we get

$$\sin \Pi(b) \sin \Pi(a) = \sin \Pi(c).$$

Equation (2) gives by changing the letters in it

$$\cos \Pi(\beta) \sin \Pi(b) = \sin \Pi(\alpha).$$

If in this equation we change β, b, α to c, α', b', we will get

$$\cos \Pi(c) \cos \Pi(\alpha) = \cos \Pi(b) \tag{3}$$

In the same way we find

$$\cos \Pi(c) \cos \Pi(\beta) = \cos \Pi(a). \tag{4}$$

The equations $(2), (3), (4)$ are related to the spherical rectangular triangle, whose sides will call in the future a, b, c and where by A, B we will denote the angles which are opposite to the sides a, b and $\frac{\pi}{2}$ will be opposite to c. In the equations mentioned above we can put a in the place of $\Pi(\beta), b$ in the place of $\Pi(c)$, c instead of $\Pi(a), \frac{\pi}{2} - A$ instead of $\Pi(a)$; B instead of $\Pi(b)$.

In such a way, the equations are

$$\sin A \sin c = \sin a,$$
$$\cos b \sin A = \cos B,. \qquad (5)$$
$$\cos a \cos b = \cos c$$

Equations (5) apply to a spherical triangle, which can be formed from a rectilinear rectangular triangle and hence whose sides cannot exceed $\frac{\pi}{2}$. Note also that if we draw through the top of the angle A the arc of the big circle perpendicular to the side b, then this arc will meet either side a or its extension in such a way that each of the two arcs from the point of intersection to the side b will be $= \frac{\pi}{2}$, and the angle between these arcs will be b.

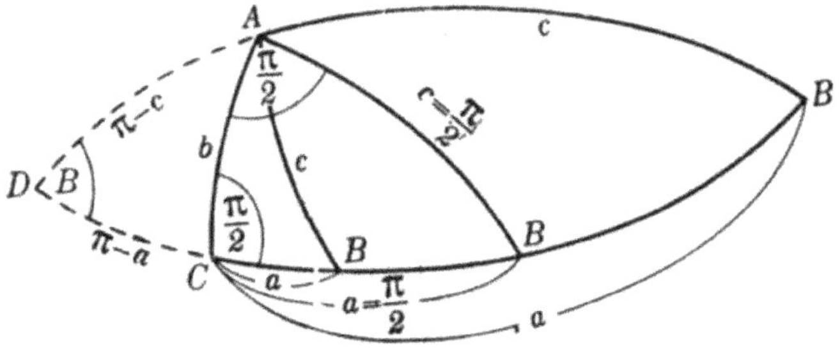

Fig. 8

After this it is not difficult to conclude, that in each rectangular spherical triangle $(ABC$, Fig. 8) if

$$c < \frac{\pi}{2}, \quad \text{that it should be} \quad a < \frac{\pi}{2} \quad \text{and} \quad A, \frac{\pi}{2},$$

if

$$c = \frac{\pi}{2}, \quad \text{then it should be} \quad a = \frac{\pi}{2} \quad \text{and} \quad A = \frac{\pi}{2},$$

at the end if

$$c > \frac{\pi}{2}, \quad \text{the it should have been} \quad a > \frac{\pi}{2} \quad \text{and} \quad A > \frac{\pi}{2};$$

from here it follows, that assuming $a > \frac{\pi}{2}$ one must also assume that $c > \frac{\pi}{2}, A > \frac{\pi}{2}$. If we extend in this case the sides a and c on the other side of side b till their intersection, then we will get another spherical rectangular triangle $[ACD]$, whose sides $\pi - a, b, \pi - c$, and the opposite angles are $\pi - A, B, \pi/2$, the triangle to which the equations (5) are applied. But equations (5) do not change their form, if we put here $\pi - a$ instead of a, $\pi - c$ instead of c and $\pi - A$ instead of A. This proves, that the equations (5) are applicable to all spherical rectangular triangles.

We go to each spherical triangle, where the sides a, b, c with opposite angles A, B, C without assuming that there is no right angle between the angles A, B, C, because equations (5) have been proven for such a case.

Draw from the top of the angle C an arc of the big circle p perpendicular to the side c; here we can have two cases: 1) Ether perpendicular p passes inside the triangle, dividing the angle C in two parts $D, C - D$ and also the side c into two parts x opposite to D, and $c - x$ opposite to $C - D$. 2) Ether perpendicular p will pass outside of the triangle adding to the angle C angle D and to the side c the arc x.

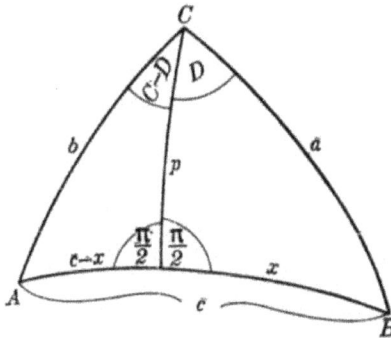

Fig. 9

In the first case (Fig. 9) the given spherical triangle will be a sum of two spherical rectangular triangles. In one of these two triangles the sides will be p, x, a with the opposite angles $B, D, \frac{\pi}{2}$ and the other side will be $p, c - x, b$ with opposite angles $A, C - D, \frac{\pi}{2}$. Applying to the first triangle equations (5), we get

$$\sin p = \sin a \sin B,$$
$$\sin x = \sin a \sin D,$$
$$\cos p \sin D = \cos B, \qquad (A)$$
$$\cos x \sin B = \cos D,$$
$$\cos a = \cos p \cos x.$$

Other triangle gives a similar result:

$$\sin p = \sin b \sin A,$$
$$\sin(c - x) = \sin b \sin(C - D),$$
$$\cos p \sin(C - D) = \cos A, \qquad (B)$$
$$\cos p \cos(c - x) = \cos b.$$

Comparing two kinds of $\sin p$ from (A) and (B) gives

$$\sin a \sin B = \sin b \sin A. \tag{6}$$

When the last from the equation (B) is divided by the last of equation (A) we get

$$\tan x = \frac{\cos b}{\cos a \sin c} - \cot c.$$

Meanwhile connecting the second with the third and the last from the equation (A) gives

$$\tan x = \tan a \cos B.$$

Comparing the two values for $\tan x$ we get the equation

$$\cos b - \cos a \cos c = \sin a \sin c \cos B. \tag{7}$$

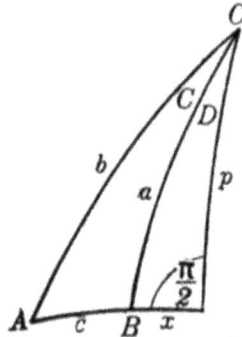

Fig. 10

If the perpendicular goes outside of the given spherical triangle, adding in such a way the arc x to the side c and the angle D to the angle C (Fig. 10), then we will have two spherical rectangular triangles. The sides of one of these two triangles will be p, x, a with opposite angles $\pi - B, D, \frac{\pi}{2}$; the sides of the other triangle will be $p, c+x, b$ with opposite angles $A, C+D, \frac{\pi}{2}$. Applying equations (5) to the first from these triangles gives:

$$\sin p = \sin a \sin B,$$
$$\sin x = \sin a \sin D,$$
$$-\cos B = \cos p \sin D, \tag{C}$$
$$\cos D = \cos x \sin B,$$
$$\cos a = \cos p \cos x.$$

The second triangle, whose sides $p, c + x, b$ with opposite angles $A, C + D, \frac{\pi}{2}$, gives the following equations

$$\sin p = \sin b \sin A,$$
$$\sin(c + x) = \sin b \sin(C + D),$$
$$\cos A = \cos p \sin(C + D), \tag{D}$$
$$\cos(C + D) = \cos(c + x) \sin A,$$
$$\cos b = \cos p \cos(c + x).$$

Comparing the two values for $\sin p$ from equations (C) and (D) again gives equation (6).

From equations (C) and (D) we find

$$\tan x = \cot c - \frac{\cos b}{\cos a \sin c}.$$

From equation (C) we find

$$\tan x = -\tan a \cos B.$$

A comparison of the two values for $\tan x$ brings us again to the equation (7), which in such a way, as equation (6), is proven for all spherical triangles in general.

Equation (7) with variable letters in it gives two more equations:

$$\cos a - \cos b \cos c = \sin b \sin c \cos A,$$

$$\cos c - \cos a \cos b = \sin a \sin b \cos C.$$

Multiplying the last equation by $\cos b$ adding the product to the first and dividing the sum by $\sin b$, we get

$$\cos a \sin b = \sin c \cos A + \sin a \cos b \cos C.$$

Putting in the following equation the value of $\sin c$

$$\sin c = \frac{\sin C}{\sin A} \sin a,$$

according to the equation (6), after dividing by $\sin a$, we get

$$\cot a \sin b = \cot A \sin C + \cos b \cos C. \tag{8}$$

Replacing here $\sin b$ in the following equation with the value

$$\sin a \frac{\sin B}{\sin A}$$

after that we multiply the equation by $\sin A$, we get

$$\cos a \sin B = \cos b \cos C \sin A + \sin C \cos A,$$

from where we get with varying the letters the following equation

$$\cos b \sin A = \cos a \cos C \sin B + \sin C \cos B.$$

The exclusion of $\cos b$ from the two last equations brings to the equation:

$$\cos a \sin B \sin C = \cos B \cos C + \cos A. \tag{9}$$

Equations (6), (7), (8), (9) are the equations which are usually given in the spherical trigonometry and which are proven with the help of the ordinary geometry. So the spherical trigonometry stays the same, if we assume, that the sum of the three angles in the rectilinear triangle always is equal to π or assume, that it is $< \pi$. It is truly remarkable with respect to the spherical trigonometry, but it does not apply to the rectilinear (ordinary) trigonometry.

4 Finding function $\Pi(x)$

6. Before deriving of the equations, which show in pangeometry the dependence of the sides and the angles of each rectilinear triangle, we will study of what should be the function of $\Pi(x)$ for each line x. Let us consider a rectilinear triangle, which sides a, b, c with opposite angles $\Pi(\alpha), \Pi(\beta), \frac{\pi}{2}$; extend the side c beyond the top of the angle $\Pi(\beta)$ and make the extension equal to β.

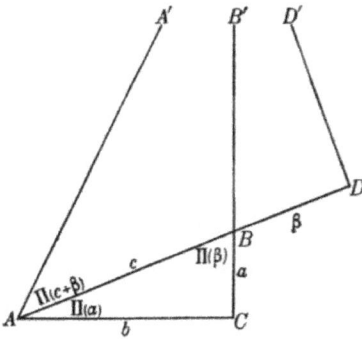

Fig. 11

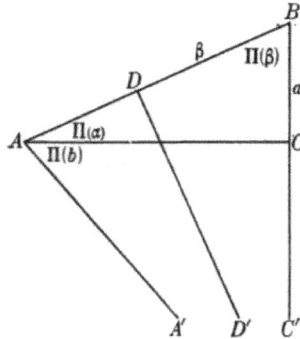

Fig. 12

At the end of line β (Fig. 11) we draw perpendicular $[DD']$, which will be parallel to the side a and with the extension $[BB']$ of this side beyond the top of the angle $\Pi(\beta)$. We draw also through the top of the angle $\Pi(\alpha)$ a parallel $[AA]'$ to the same extension of the side a. The angle, which this straight line makes with the side c will be $\Pi(c + \beta)$, the angle which the straight line makes with b, will be $\Pi(b)$. In such a way we will get the following equation

$$\Pi(b) = \Pi(c + \beta) + \Pi(\alpha). \tag{II}$$

If we draw the line β from the top of the angle $\Pi(\beta)$ to the side c (Fig. 12) and to the line β draw at the end $[D]$ perpendicular $[DD'][$to$]$ the side of the angle $\Pi(\beta)$, then this straight line will be parallel to the extension a [i.e., to CC'] beyond the top of the right angle. We draw through the top of the angle $\Pi(\alpha)$ parallel to $[AA']$ to the last perpendicular, which

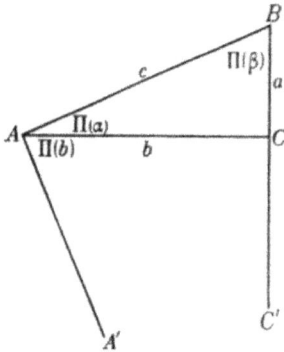

Fig. 13

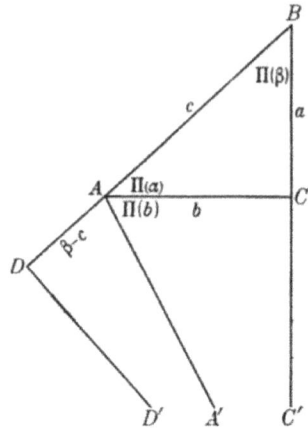

Fig. 14

will be also parallel to the second extension $[CC']$ of the side a, making the angle $\Pi(c - \beta)$ with side c and $\Pi(b)$ with side b; hence we get

$$\Pi(b) = \Pi(c - \beta) - \Pi(\alpha). \tag{Π'}$$

It is easy to see, that this equation holds not only for c greater than β, but also for $c = \beta$ and $c < \beta$. Indeed, if $c = \beta$ (Fig. 13), then $\Pi(c - \beta) = \Pi(0) = \frac{\pi}{2}$; on the other side, the perpendicular $[AA']$ to c through the top of the angle $\Pi(\alpha)$ is made parallel to a, from where it follows, that $\Pi(b) = \frac{\pi}{2} - \Pi(\alpha)$, which is in accordance with our equations. If $c < \beta$ (Fig. 14), then the end $[D]$ of the line β will go beyond the top of the angle $\Pi(\alpha)$ at a distance $\beta - c$ from the top. The perpendicular $[DD']$ to β at the end of $[D]$ of this line will be parallel to a, $[BC']$ and with the straight line $[AA']$, drawn through the top of the angle $\Pi(\alpha)$ parallel to a; from where it follows, that two adjacent angles whose this parallel line makes with the side c are astute equal to $\Pi(\beta - c)$, and obtuse equal to $\Pi(\alpha) + \Pi(b)$. The sum of two adjacent angles makes π, so

$$\Pi(\beta - c) + \Pi(\alpha) + \Pi(b) = \pi$$

or

$$\Pi(b) = \pi - \Pi(\beta - c) - \Pi(\alpha).$$

But according to the definition of function $\Pi(x)$

$$\pi - \Pi(\beta - c) = \Pi(c - \beta),$$

after that

$$\Pi(b) = \Pi(c - \beta) - \Pi(\alpha),$$

the same equation as the one found above and which in such a way was proven for all cases.

Two equations (Π) and (Π') can be replaced the following equations:

$$\Pi(b) = \frac{1}{2}\Pi(c+\beta) + \frac{1}{2}\Pi(c-\beta),$$

$$\Pi(\alpha) = \frac{1}{2}\Pi(c+\beta) - \frac{1}{2}\Pi(c-\beta);$$

but the equation (3) gives

$$\cos\Pi(c) = \frac{\cos\Pi(b)}{\cos\Pi(\alpha)}.$$

Inserting in this equation in the place of $\Pi(b)$ and $\Pi(\alpha)$ their values, we find:

$$\cos\Pi(c) = \frac{\cos\{\frac{1}{2}\Pi(c+\beta) + \frac{1}{2}\Pi(c-\beta)\}}{\cos\{\frac{1}{2}\Pi(c-\beta) - \frac{1}{2}\Pi(c+\beta)\}}.$$

From this equation we derive the following:

$$\tan^2 \frac{1}{2}\Pi(c) = \tan\frac{1}{2}\Pi(c-\beta)\tan\frac{1}{2}\Pi(c+\beta).$$

Because the lines c, β can change independently from each other in every rectilinear rectangular triangle, then putting gradually in the last equations $c = \beta, c = 2\beta, c = 3\beta, \ldots, c = n\beta$, we derive from these equations for each positive integer n

$$\tan^n \frac{1}{2}\Pi(c) = \tan\frac{1}{2}\Pi(nc).$$

It is easy to prove that this equation holds for numbers n negative or fractions. From here it follows, that taking for the unit of straight line such which gives:

$$\tan\frac{1}{2}\Pi(1) = e^{-1},$$

where e is base of Neper logarithms, then we get for each line x

$$\tan\frac{1}{2}\Pi(x) = e^{-x}.$$

This expression gives $\Pi(x) = \frac{1}{2}\pi$ for $x = 0$, and $\Pi(x) = 0$ for $x = \infty$, $\Pi(x) = \pi$ for $x = -\infty$, in accordance with what was accepted and proven above.

7. The value, found for $\tan\frac{1}{2}\Pi(x)$ gives for each line x

$$\sin\Pi(x) = \frac{2}{e^x + e^{-x}}, \qquad \cos\Pi(x) = \frac{e^x - e^{-x}}{e^x + e^{-x}}$$

and for two arbitrary lines x, y

$$\sin \Pi(x + y) = \frac{\sin \Pi(x) \sin \Pi(y)}{1 + \cos \Pi(x) \cos \Pi(y)},$$

$$\sin \Pi(x - y) = \frac{\sin \Pi(x) \sin \Pi(y)}{1 - \cos \Pi(x) \cos \Pi(y)},$$

$$\cos \Pi(x + y) = \frac{\cos \Pi(x) + \cos \Pi(y)}{1 + \cos \Pi(x) \cos \Pi(y)},$$

$$\cos \Pi(x - y) = \frac{\cos \Pi(x) - \cos \Pi(y)}{1 - \cos \Pi(x) \cos \Pi(y)},$$

$$\tan \Pi(x + y) = \frac{\sin \Pi(x) \sin \Pi(y)}{\cos \Pi(x) + \cos \Pi(y)}.$$

5 EQUATIONS CONNECTING SIDES AND ANGLES OF ANY TRIANGLE

8. Equations (2), (3), (4) found for spherical rectangular triangles also apply to the rectilinear rectangular triangle, whose sides a, b, c have opposite angles $\Pi(\alpha), \Pi(\beta), \frac{\pi}{2}$. So putting A in the place of $\Pi(\alpha)$, B in the place of $\Pi(\beta)$, we will get for each rectilinear rectangular triangle whose sides are a, b, c and where A is the angle opposite to a, B opposite to b and $\frac{\pi}{2}$ angle opposite to c the following equations:

$$\sin \Pi(a) \cos A = \sin B,$$
$$\cos \Pi(c) \cos A = \cos \Pi(b), \qquad (10)$$
$$\cos \Pi(c) \cos B = \cos \Pi(a).$$

To these equations we add the following, which also had been proven above:

$$\sin \Pi(a) \sin \Pi(b) = \sin \Pi(c). \qquad (11)$$

The first from the equations (10) with varying letters in it can be written in this way as well:

$$\sin \Pi(b) \cos B = \sin A.$$

Inserting here the value of $\cos B$ from the third equation (10) we get:

$$\sin \Pi(b) \cos \Pi(a) = \sin A \cos \Pi(c).$$

Excluding from this equation $\sin \Pi(b)$ with the help of equation (11) we get:

$$\tan \Pi(c) = \sin A \tan \Pi(a). \qquad (12)$$

9. Let now a, b, c be sides in general to some rectilinear triangle, after that A, B, C angles which are opposite to these sides. Draw perpendicular p from the top of the angle C on the side c. If p is falling inside the triangle (Fig. 15) and divides the angle C into two angles D and $C - D$ and the side c into two parts: x opposite to D, $c - x$ opposite to $C - D$, then there will be formed two rectilinear rectangular triangles.

135

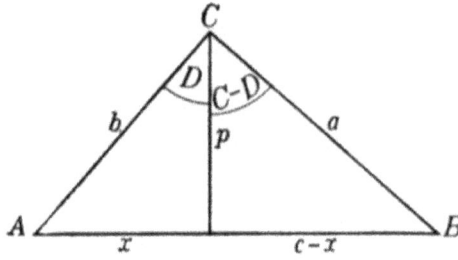

Fig. 15

In one side will be p, x, b with opposite angles $A, D, \frac{\pi}{2}$; on the other side will be $p, c - x, a$ with opposite angles $B, C - D, \frac{\pi}{2}$. The use of equation (12) to the first from these two triangles gives:

$$\tan \Pi(b) = \sin A \tan \Pi(p).$$

The second of these two triangles gives in the same way the following equation:

$$\tan \Pi(a) = \sin B \tan \Pi(p);$$

from where we conclude:

$$\sin A \tan \Pi(a) = \sin B \tan \Pi(b). \tag{13}$$

The application of the equations (10) and (11) to the first of two triangles gives:

$$\cos \Pi(b) \cos A = \cos \Pi(x),$$
$$\sin \Pi(x) \sin \Pi(p) = \sin \Pi(b).$$

The second triangle gives:

$$\sin \Pi(p) \sin \Pi(c - x) = \sin \Pi(a).$$

Inserting in this last equation in the place of $\sin \Pi(c - x)$ its value, taken from general formula, that we have found above for $\Pi(x - y)$, we will get:

$$\frac{\sin \Pi(a)}{\sin \Pi(p)} = \frac{\sin \Pi(c) \sin \Pi(x)}{1 - \cos \Pi(c) \cos \Pi(x)};$$

from where we conclude, inserting

$$\sin \Pi(p) = \frac{\sin \Pi(b)}{\sin \Pi(x)},$$

$$\cos \Pi(x) = \cos \Pi(b) \cos A,$$

the following equation:

$$1 - \cos \Pi(b)(\cos \Pi(c) \cos A = \frac{\sin \Pi(b) \sin \Pi(c)}{\sin \Pi(a)}. \tag{14}$$

Equations (13), (14) are verified themselves for $A = \frac{\pi}{2}$, when the perpendicular·p coincide with the side b, because in this case equation (13) changes into equation (12), and equation (14) goes into equation (11), equations, proven for every rectilinear right triangle.

If perpendicular p falls out of the triangle on the extension of the side c (Fig. 16), adding the line x to the side c and the angle D to the angle C, then we form two rectangular triangles. The sides of one of them will be p, x, b with the opposite angles $(\pi - A), D, \frac{\pi}{2}$, the sides of the other will be $p, c + x, a$ with opposite angles $B, C + D, \frac{\pi}{2}$. Using equation (12) in the first from these two triangles gives:

$$\tan \Pi(b) = \sin A \tan \Pi(p).$$

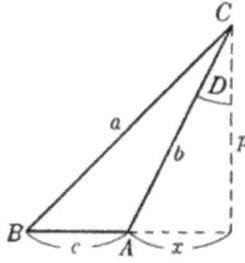

Fig. 16

From the other triangle we derive in the same way the following equation:

$$\tan \Pi(a) = \sin B \tan \Pi(p).$$

Excluding $\tan \Pi(p)$ from the two last equations we again find equation (13). Using equations (10) and (11) in the first of two triangles gives:

$$- \cos \Pi(b) \cos A = \cos \Pi(x),$$

$$\sin \Pi(b) = \sin \Pi(x) \sin \Pi(p).$$

From the second triangle we find in the same way:

$$\sin \Pi(a) = \sin \Pi(p) \sin \Pi(c + x).$$

Replacing in this equation $\sin \Pi(c+x)$ with its value, taken from the general formula, found above for $\sin \Pi(x + y)$, we get:

$$\frac{\sin \Pi(a)}{\sin \Pi(p)} = \frac{\sin \Pi(c) \sin \Pi(x)}{1 + \cos \Pi(c) \cos \Pi(x)}.$$

Inserting in the following equation here

$$\sin \Pi(p) = \frac{\sin \Pi(b)}{\sin \Pi(x)},$$

$$\cos \Pi(x) = -\cos \Pi(b) \cos A,$$

we get:

$$\frac{\sin \Pi(a)}{\sin \Pi(b)} = \frac{\sin \Pi(c)}{1 - \cos \Pi(b) \cos \Pi(c) \cos A},$$

which equation is identical to the equation (14).

In such a way, the equations (13), (14) are proven for each rectilinear right triangle.

Equation (14) with varied letters gives:

$$1 - \cos \Pi(c) \cos \Pi(a) \cos B = \frac{\sin \Pi(c) \sin \Pi(a)}{\sin \Pi(b)}.$$

Multiplying this equation term by term by the equation (14), we get:

$$1 - \cos \Pi(c) \cos \Pi(a) \cos B - \cos \Pi(b) \cos \Pi(c) \cos A +$$
$$+ \cos \Pi(a) \cos \Pi(b) \cos^2 \Pi(c) \cos A \cos B = \sin^2 \Pi(c)$$

or

$$\cos^2 \Pi(c) - \cos \Pi(c) \cos \Pi(a) \cos B - \cos \Pi(b) \cos \Pi(c) \cos A +$$
$$+ \cos \Pi(a) \cos \Pi(b) \cos^2 \Pi(c) \cos A \cos B = 0.$$

Subtracting from this equation the common factor $\cos \Pi(c)$ we get:

$$\cos \Pi(c) - \cos \Pi(a) \cos B - \cos \Pi(b) \cos A +$$
$$+ \cos \Pi(a) \cos \Pi(b) \cos \Pi(c) \cos A \cos B = 0.$$

In the same way we find:

$$\cos \Pi(a) - \cos \Pi(b) \cos C - \cos \Pi(c) \cos B +$$
$$+ \cos \Pi(a) \cos \Pi(b) \cos \Pi(c) \cos B \cos C = 0.$$

Multiplying this equation by $\cos A$ and subtracting product from product of the previous equation by $\cos C$, we get:

$$\cos \Pi(a)\{\cos A + \cos B \cos C\} = \cos \Pi(c)\{\cos C + \cos A \cos B\}.$$

Squaring both parts of this equation and after dividing in $\cos^2 \Pi(c)$ it takes the following form:

$$\frac{\cos^2 \Pi(a)}{\cos^2 \Pi(c)}\{\cos A + \cos B \cos C\}^2 = \{\cos C + \cos A \cos B\}^2.$$

Meanwhile equation (13) gives:

$$\frac{1}{\cos^2 \Pi(c)} = 1 + \frac{\sin^2 A}{\sin^2 C} \tan^2 \Pi(a).$$

If we put in the equation before the last instead of $\frac{1}{\cos^2 \Pi(c)}$ its value from the last equation, then we get:

$$\cos^2 \Pi(a) + \frac{\sin^2 A}{\sin^2 C} \sin^2 \Pi(a) = \left\{ \frac{\cos C + \cos A \cos B}{\cos A + \cos B \cos C} \right\},$$

after:

$$\sin^2 \Pi(a) \left\{ 1 - \frac{\sin^2 A}{\sin^2 C} \right\} = \frac{\sin^2 B (\sin^2 C - \sin^2 A)}{(\cos A + \cos B \cos C)^2}.$$

Dividing this equation by $\sin^2 C - \sin^2 A$ and taking square root, we find;

$$\sin \Pi(a) = \frac{\sin B \sin C}{\cos A + \cos B \cos C}$$

without reciprocity in the signs, because the terms of the last equation are both positive. Indeed $\Pi(a) < \frac{\pi}{2}$; $B < \pi$; $C < \pi$, from where follows, that sinuses of these angles are positive, after we have:

$$\cos A + \cos(B + C) = 2 \cos \frac{1}{2}(A + B + C) \cos \frac{1}{2}(B + C - A),$$

but $A + B + C < \pi$, hence $\cos \frac{1}{2}(A + B + C)$ is positive, as $\cos \frac{1}{2}(B + C - A)$; adding to both terms of the last equation the positive number $\sin B \sin B$, we find: $\cos A + \cos B \cos C > 0$. Therefore in each rectilinear triangle:

$$\cos A + \cos B \cos C = \frac{\sin B \sin C}{\sin \Pi(a)} \qquad (15)$$

Multiplying equation (14) term by terms of the following equation, which comes from the above by varying letters

$$1 - \cos \Pi(a) \cos \Pi(b) \cos C = \frac{\sin \Pi(a) \sin \Pi(b)}{\sin \Pi(c)} \qquad (16)$$

gives

$$\{1 - \cos \Pi(a) \cos \Pi(b) \cos C\}\{1 - \cos \Pi(b) \cos \Pi(c) \cos A\} = \sin^2 \Pi(b),$$

which, after the multiplication looks like that:

$$\cos^2 \Pi(b) - \cos \Pi(a) \cos \Pi(b) \cos C - \cos \Pi(b) \cos \Pi(c) \cos +$$
$$A + \cos^2 \Pi(b) \cos \Pi(a) \cos \Pi(c) \cos A \cos C = 0$$

or, dividing by $\cos \Pi(b)$,

$$\cos \Pi(b) - \cos \Pi(a) \cos C - \cos \Pi(c) \cos A +$$
$$+ \cos \Pi(a) \cos \Pi(b) \cos \Pi(c) \cos A \cos C = 0 \qquad (17)$$

But we find according to the equation (13),

$$\cos \Pi(c) = \frac{\sin \Pi(c) \sin C}{\sin A} \cot \Pi(a).$$

in this equation we can insert instead of $\sin \Pi(c)$ its value, taken from the equation (16); we will get:

$$\cos \Pi(c) = \frac{\sin \Pi(b) \cos \Pi(a) \sin C}{\{1 - \cos \Pi(a) \cos \Pi(b) \cos C\} \sin A}.$$

Putting the value of $\cos \Pi(c)$ in the equation (17), we get:

$$\cot A \sin C \sin \Pi(b) + \cos C = \frac{\cos \Pi(b)}{\cos \Pi(a)}. \tag{18}$$

Let us connect the equation (13), (14), (15) and (18) which show dependence of the angles and the sides of each rectilinear triangle, in such a way that can be facilitate their application:

$$\sin A \tan \Pi(a) = \sin B \tan \Pi(b),$$

$$1 - \cos \Pi(b) \cos \Pi(c) \cos A = \frac{\sin \Pi(b) \sin \Pi(c)}{\sin \Pi(a)},$$

$$\cos A + \cos B \cos C = \frac{\sin B \sin C}{\sin \Pi(a)}, \tag{19}$$

$$\cot A \sin C \sin \Pi(b) + \cos C = \frac{\cos \Pi(b)}{\cos \Pi(a)}.$$

With the help of these equations pangeometry goes into Analytical Geometry and in such a way forms the special complete geometrical theory. Equations (19) are used for presenting the equations of curved lines between the coordinates and their points, also for calculation of lengths and areas of curved lines, surfaces and the volumes of bodies as I showed in my article in the journal University of Kazan of 1829 year.

10. It was shown above, that pangeometry goes into ordinary geometry with the assumption that the lines are extremely small. Now we can believe that this is true. For each extremely small line x can be used approximative values:

$$\cot \Pi(x) = x,$$

$$\sin \Pi(x) = 1 - \frac{1}{2}x^2,$$

$$\cos \Pi(x) = x.$$

If we consider the side in the triangle as infinitesimal of the first order, and if we neglect infinitesimal quantities of the second and higher order, then equations (19) after insertion of approximate values $\sin \Pi(a)$, $\sin \Pi(b)$ and so on will look as the following equations:

$$b \sin a = a \sin B,$$

$$a^2 = b^2 + c^2 - 2bc \cos A,$$

$$\cos A + \cos(B + C) = 0,$$

$$a \sin(A + C) = b \sin A.$$

The first two of these equations are well known in the ordinary geometry. The last two equations give

$$A + B + C = \pi.$$

6 FOUNDATIONS OF ANALITICAL GEOMETRY. LENGTH OF CIRCUMFERENCE AND ARC OF LIMITING CIRCE

11. In order to give an example how curved lines are defined with the help of the coordinates of their points, we call y a perpendicular, drawn from a point on the circle (Fig.17) to one of its fixed diameters, whose magnitude we call by $2r$, hence under r we understand the radius of the circle.

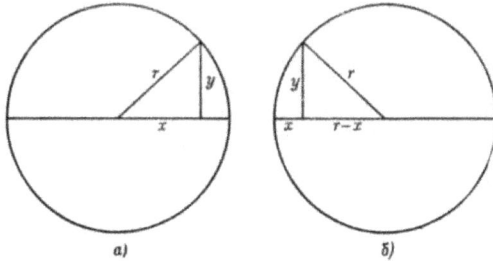

Fig. 17

We call also x the part of this diameter from the center to the perpendicular y (Fig. 17, a). Applying equation (11) to the right triangle, with sides x, y, r gives

$$\sin \Pi(x) \sin \Pi(y) = \sin \Pi(r), \tag{20}$$

which forms *the equation of a circle* between the right perpendicular coordinates x, y.

If we consider x from the end of the diameter of the circle (Fig. 17, b), then the equation (20) will look like that:

$$\sin \Pi(r - x) \sin \Pi(y) = \sin \Pi(r) \tag{20'}$$

or equal to:

$$2(e^r + e^{-r}) = (e^{r-x} + e^{-r+x})(e^y + e^{-y}).$$

If we divide this equation by e^r and if after we take $r = \infty$, then we will get the *equation for the limiting circle*

$$2 = (e^y + e^{-y})e^{-x}$$

or

$$\sin \Pi(y) = \tan \frac{1}{2}\Pi(x). \tag{20a}$$

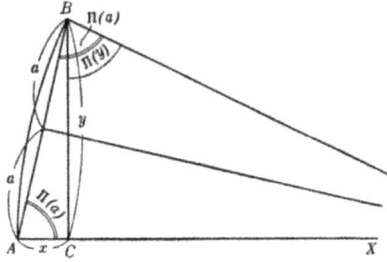

Fig. 18

From the definition of the limiting circle it follows, that two axes of the limiting circle, drawn through the ends of one chord, can be inclined to this chord under equal angles, which property can be taken for the definition of the limiting circle. From here we can derive the equation of this curve, considering triangle (Fig.18) with sides x, y and a chord $2a$ of the limiting circle. The angles of this triangle will be : $\Pi(a) - \Pi(y)$ opposite to x, $\Pi(a)$ opposite to y and $\frac{\pi}{2}$ opposite to $2a$. According to the equations (10), (11), in this triangle we will have:

$$\sin \Pi(x) \sin \Pi(y) = \sin \Pi(2a),$$
$$\sin \Pi(x) \cos\{\Pi(a) - \Pi(y)\} = \sin \Pi(a),$$
$$\sin \Pi(y) \cos \Pi(a) = \sin\{\Pi(a) - \Pi(y)\}.$$

The last equation gives:

$$2 \tan \Pi(y) = \tan \Pi(a), \tag{21}$$

the first equation can be written the following way:

$$\sin \Pi(x) \sin \Pi(y) = \frac{\sin^2 \Pi(a)}{1 + \cos^2 \Pi(a)}.$$

Inserting in this equation instead of $\sin^2 \Pi(a), 1 + \cos^2 \Pi(a)$ their values, expressed by $\tan^2 \Pi(a)$ and adding here $\tan^2 \Pi(a)$ from the equation (21), we find:

$$\sin \Pi(x) \sin \Pi(y) = \frac{2 \tan^2 \Pi(y)}{1 + 2 \tan^2 \Pi(y)},$$

after that:

$$\sin \Pi(x) = \frac{2 \sin \Pi(y)}{1 + \sin^2 \Pi(y)},$$

from here we derive:

$$2\cos^2 \frac{1}{2}\Pi(x') = \frac{\{1 + \sin \Pi(y)\}^2}{1 + \sin^2 \Pi(y)}, \quad 2\sin^2 \frac{1}{2}\Pi(x') = \frac{\{1 - \sin \Pi(y)\}^2}{1 = \sin^2 \Pi(y)}.$$

Dividing the last from these equations by the equation before the last and taking square root, we get:

$$\tan \frac{1}{2}\Pi(x') = \frac{1 - \sin \Pi(y)}{1 + \sin \Pi(y)};$$

from here

$$\sin \Pi(y) = \frac{1 - \tan \frac{1}{2}\Pi(x')}{1 + \tan \frac{1}{2}\Pi(x')}.$$

The second part of this equation can be presented in the following way:

$$\frac{\cos \frac{1}{2}\Pi(x') - \sin \frac{1}{2}\Pi(x')}{\cos \frac{1}{2}\Pi(x') + \sin \frac{1}{2}\Pi(x')} = \frac{\sin\{\frac{1}{4}\pi - \frac{1}{2}\Pi(x')\}}{\cos\{\frac{1}{4}\pi - \frac{1}{2}\Pi(x')\}} = \frac{\sin \frac{1}{2}\Pi(x)}{\cos \frac{1}{2}\Pi(x)} = \tan \frac{1}{2}\Pi(x).$$

and hence

$$\sin \Pi(y) = \tan \frac{1}{2}\Pi(x), \tag{20a}$$

as it was found above.

12. In order that we give an example how to calculate the length of a curved line, we look for *the length of the circumferences of the circle*, which radius is r. Let us introduce two radii, whose angle at the center is $\frac{2\pi}{n}$, where n is a whole number. Let us draw from the end one of the radii a perpendicular p to the other radius (Fig. 19). The product np will be as closer to the length of the circumference then the number n is greater.

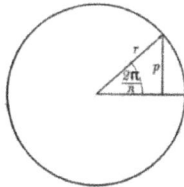

Fig. 19

The righ triangle, where p is one of the sides, r is hypotenuse, and $\frac{2\pi}{n}$ the angle opposite to p, gives (equation 13):

$$\sin \frac{2\pi}{n} \tan \Pi(p) = \tan \Pi(r).$$

But it is known, that

$$\{n \sin \frac{2\pi}{n}\} = 2\pi \quad \text{for} \quad n = \infty,$$

while

$$\frac{\tan \Pi(p)}{n} = \frac{2}{n(e^p - e^{-p})}$$

and

$$n(e^p - e^{-p}) = 2np$$

with as greater an accuracy as the number n is greater, so p is smaller. After that the length of the circumference [radius] r

$$= np = 2\pi \cot \Pi(r)$$

or the length of the circumference [radius] r [equal] $\pi(e^r - e^{-r})$; this gives for r an extremely small number of the length of the circumference $r = 2\pi r$, the same as in the ordinary geometry.

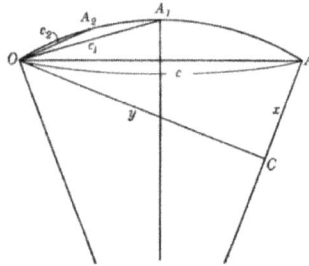

Fig. 20

We define *the arc* $s[OA]$ *of the limiting circle* (Fig. 20) through the coordinates: y – a perpendicular to $[OC]$, drawn from one end $[O]$, drawn from the end $[O]$ of the arc s to the axis drawn through the other end, and x part of $[AC]$ of this axis between the top and the perpendicular y. Let c be the chord $[OA]$ of the arc s, let also c_1, c_2, c_3 and so on be the chords of the arcs $\frac{1}{2}s, \frac{1}{2^2}s, \frac{1}{2^3}s, \ldots$. We proved above (equation 21), that:

$$\cot \Pi(y) = 2 \cot \Pi\left(\frac{1}{2}c\right).$$

In a similar way we will have the following equations:

$$\cot \Pi\left(\frac{1}{2}c\right) = 2 \cot \Pi\left(\frac{1}{2}c_1\right),$$

$$\cot \Pi\left(\frac{1}{2}c_1\right) = 2 \cot \Pi\left(\frac{1}{2}c_2\right),$$

$$\cot \Pi\left(\frac{1}{2}c_2\right) = 2 \cot \Pi\left(\frac{1}{2}c_3\right)$$

in general for each whole and positive number n

$$\cot \Pi\left(\frac{1}{2}c_{n-1}\right) = 2 \cot \Pi\left(\frac{1}{2}c_n\right).$$

From here we conclude:

$$\cot \Pi(y) = 2^{n+1} \cot \Pi\left(\frac{1}{2}c_n\right).$$

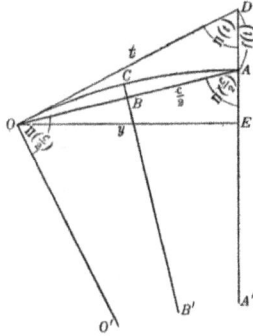

Fig. 21

If n is very great hence c_n is very small, we get

$$2^{n+1} \cot \Pi\left(\frac{1}{2}c_n\right) = 2^n c_n.$$

But

$$2^n c_n = s \quad \text{for} \quad n = \infty,$$

from here it follows, that

$$s = \cot \Pi(y). \tag{22}$$

Determine furthermore the arc s of the limiting circle (Fig. 21) with the help of t, parts of the tangent $[OD]$, drawn to the limiting circle in the apex $[O]$ of the axes $[OO']$, through the end of the arc s between the touching points and point $[D]$ the intersection of the tangent with the axis through the other end of the arc s, i.e., deifying function $L(t)$. In the triangle where the sides $c, t, f(t)$ with opposite angles $\Pi(t), \pi - \Pi(\frac{1}{2}c)$, $\frac{1}{2}\pi - \Pi(\frac{1}{2}c)$, we find using equation (13)

$$\sin \Pi(t) \tan \Pi(c) = \sin \Pi\left(\frac{1}{2}c\right) \tan \Pi(t);$$

but we have seen (equation 21), that

$$\tan \Pi\left(\frac{1}{2}c\right) = 2 \tan \Pi(y),$$

to this we add a remark, that

$$\tan \Pi(c) = \frac{\sin^2 \Pi\left(\frac{1}{2}c\right)}{2\cos \Pi\left(\frac{1}{2}c\right)}.$$

After that

$$\cos \Pi(t) = 2 \cot \Pi\left(\frac{1}{2}c\right),$$

i.e., as a result of the equation (22)

$$\cos \Pi(t) = s = L(t). \tag{22a}$$

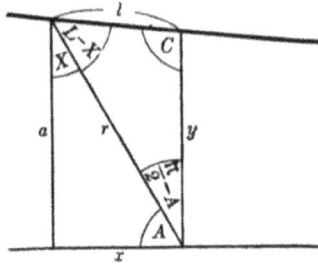

Fig. 22

13. *The equation of a straight line* is sufficiently difficult, if this equation should have been common and is relevant for each position of the line concerning the axes of the coordinates. Let us draw from a given point on a given straight line perpendicular a (Fig. 22) to the axis x and call it L angle, which this perpendicular makes with the straight line.

Further we call y the perpendicular, drawn to the axis x from another point of a given straight line. Let l be a distance from first point to the second, and let x be part of the axis x between the two perpendiculars. We introduce the straight line r from the top a till the end y and get two triangles; 1), right, whose sides a, x, r with opposite angles $A, X, \frac{\pi}{2}$, 2) with sides y, r, l with the opposite angles $L - X, C, \frac{\pi}{2} - A$. Applying equations (10), (11) to the first of these triangles gives:

$$\sin \Pi(x) \sin \Pi(a) = \sin \Pi(r),$$
$$\sin \Pi(x) \cos X = \sin A,$$
$$\sin \Pi(a) \cos A = \sin X,$$
$$\cos \Pi(r) \cos A = \cos \Pi(x),$$
$$\cos \Pi(r) \cos X = \cos \Pi(a).$$

From the above equations we derive the following:

$$\tan A = \tan \Pi(x) \cos \Pi(a),$$
$$\tan \Pi(r) = \tan \Pi(x) \sin \Pi(a) \cos A,$$
$$\tan X = \tan \Pi(a) \cos \Pi(x), \tag{22b}$$
$$\cos \Pi(x) = \cos \Pi(r) \cos A,$$
$$\sin X = \sin \Pi(a) \cos A.$$

Adding the last of equations (19) to the second triangle gives:

$$\cot(L - X)\cos A \sin \Pi(r) + \sin A = \frac{\cos \Pi(r)}{\cos \Pi(y)},$$

from where it follows, that:

$$\cos \Pi(y) = \frac{\cos \Pi(r)}{\cot(L - X)\cos A \sin \Pi(r) + \sin A},$$

$$\cos \Pi(y) = \frac{\cos \Pi(r)(\tan L - \tan X)}{\{1 + \tan L \tan X\}\cos A \sin \Pi(a) \sin \Pi(x) + \sin A\{\tan L - \tan X\}}.$$

Inserting in this equation in the place of $\tan X$ its value, we get

$$\cos \Pi(y) =$$
$$= \frac{\cos \Pi(r)\{\tan L - \tan \Pi(a)\cos \Pi(x)\}}{\{1 + \tan L \tan \Pi(a)\cos \Pi(x)\}} \times$$
$$\times \frac{1}{\cos A \sin \Pi(a) \sin \Pi(x) + \sin A\{\tan L - \tan \Pi(a)\cos \Pi(x)\}}.$$

Inserting in the above equation, instead of $\cos \Pi(r)$ its value, we find that:

$$\cos \Pi(y) = \frac{\cos \Pi(a)}{\sin \Pi(x)} - \sin \Pi(a)\cot \Pi(x)\cot L. \qquad (23)$$

If the straight line is parallel to the axis x, then $L = \Pi(a)$ and the equation (28) will become:

$$\cos \Pi(y) = \frac{\cos \Pi(a)}{\sin \Pi(x)} - \frac{\cos \Pi(a)}{\tan \Pi(x)} \qquad (23a)$$

or

$$\cos \Pi(y) = \cos \Pi(a)e^{-x}. \qquad (24)$$

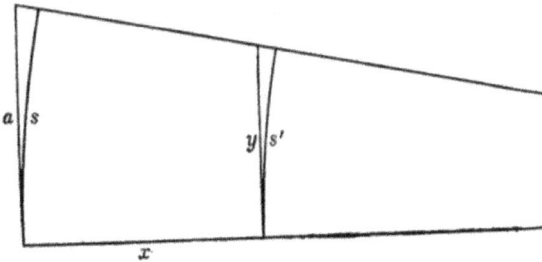

Fig. 23

If we call s, s' the length of the two arcs of the limiting circle, between the axis x and the straight line parallel to it (Fig. 23), plus from the first arc s draw tangent to a on the base, and the second tangent to y on the base, then we get the result according to the proven above

$$s = \cos \Pi(a)$$

$$s' = \cos \Pi(y),$$

after that

$$s' = se^{-x},$$

where x is a distance between two arcs s and s'. This equation shows, that the constant E, introduced above, in order to denote the content of the two arcs of the limiting circle between the two parallel, arcs, whose distance is equal to unity, is equal to e, i.e., the base of Neper logarithms.

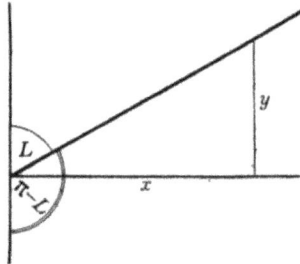

Fig. 24

If in the equation (23) $a = 0$ and if put $\pi - L$ in the place of L then we get:

$$\cos \Pi(y) = \cot \Pi(x) \cot L; \tag{24a}$$

this equation belongs to the straight line, which passes through the beginning of coordinate x, forming the angle $\frac{\pi}{2} - L$ with the axis x (Fig. 24), which is in agreement with equation (10).

7 EQUATIONS RELATING THE ELEMENTS OF A QUADRANGLE WITH THREE RIGHT ANGLES AND THEIR APPLICATION

14. Consider a quadrangle (Fig. 25) whose two sides a, y are perpendicular to the third side x. Let c be the fourth side, and φ be the angle between a and c, when the angle between c and y is right. Draw a diagonal r from the apex of the angle φ to the top of the opposite right angle. This diagonal divides the quadrangle into two right triangles; the sides of one of these triangles will be a, x, r with the opposite angles $A, X, \frac{\pi}{2}$, the sides of the other triangle y, c, r with the opposite angles $\varphi - X, \frac{\pi}{2} - A, \frac{\pi}{2}$.

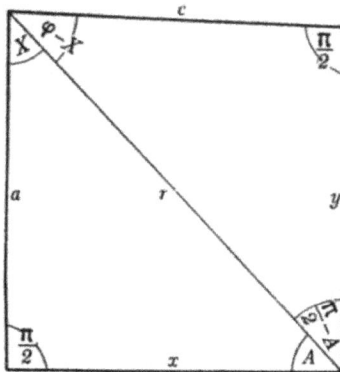

Fig. 25

Using the equation (10), (11), (13) in the first of these triangles gives:

$$
\begin{aligned}
\sin \Pi(r) &= \sin \Pi(a) \sin \Pi(x), \\
\sin A \tan \Pi(a) &= \sin X \tan \Pi(x), \\
\cos \Pi(r) \cos A &= \cos \Pi(x), \\
\cos \Pi(r) \cos X &= \cos \Pi(a);
\end{aligned}
\tag{G}
$$

151

the second triangle gives in the same manner the following equations:

$$\sin \Pi(y) \sin \Pi(c) = \sin \Pi(r),$$
$$\sin \Pi(y) \cos(\varphi - X) = \cos A,$$
$$\cos \Pi(r) \cos(\varphi - X) = \cos \Pi(c),$$
$$\cos \Pi(r) \sin A = \cos \Pi(y).$$

(H)

Equation (12), applied to the first triangle, gives:

$$\tan \Pi(r) = \sin X \tan \Pi(x)$$
$$\tan \Pi(r) = \sin A \tan \Pi(a).$$

(K)

When we apply this equation to the second triangle we get:

$$\tan \Pi(r) = \sin(\varphi - X) \tan \Pi(y),$$
$$\tan \Pi(r) = \cos A \tan \Pi(c).$$

(L)

Inserting in the second of the equations K in the place of $\sin \Pi(r)$ its value, taken from the equation (G), find:

$$\cos \Pi(r) = \frac{\sin \Pi(x) \cos \Pi(a)}{\sin A}$$

Inserting this value $\cos \Pi(r)$ in the last equation (H) we get:

$$\cos \Pi(y) = \sin \Pi(x) \cos \Pi(a).$$

(25)

Dividing the last of the equations (H) by the third of the equations (G), we get:

$$\tan A = \frac{\cos \Pi(y)}{\cos \Pi(x)};$$

inserting in this equation in the place of $\cos \Pi(y)$ its value from (25); we have:

$$\tan A = \tan \Pi(x) \cos \Pi(a).$$

Dividing of the second of equations (G) by the last of these equations term by term gives:

$$\frac{\tan X \tan \Pi(x)}{\cos \Pi(r)} = \frac{\sin A \tan \Pi(a)}{\cos \Pi(a)}.$$

Inserting in this equation at the place of $\sin A$ its value, taken from the last in the equations (H), we get:

$$\tan X = \frac{\cos \Pi(y) \tan \Pi(a)}{\cos \Pi(a)} \cot \Pi(x).$$

Replacing in this equation $\cos \Pi(y)$ by its value, found above we get:

$$\tan X = \cos \Pi(x) \tan \Pi(a).$$

Adding the second of the equation (H) to the first of the equation (L) gives also:

$$\tan(\varphi - X)\frac{\tan \Pi(y)}{\sin \Pi(y)} = \frac{\tan \Pi(r)}{\cos A}$$

or

$$\tan(\varphi - X) = \frac{\cos \Pi(y) \tan \Pi(r)}{\cos A},$$

and if we insert the value of $\tan \Pi(r)$ from the second equation into the equation (K), then we get:

$$\tan(\varphi - X) = \tan A \tan \Pi(a) \cos \Pi(y).$$

This equation, which we get when replace $\tan A$ and $\tan X$ with their values, found above will look like the following:

$$\tan \varphi = \frac{\tan \Pi(a)}{\cos \Pi(a)}. \tag{26}$$

This equation shows that using of x is always possible, as long as the angle φ is greater then $\Pi(a)$ and smaller then $\frac{\pi}{2}$, or

$$\pi - \varphi > \Pi(a); \quad \Pi - \varphi < \frac{\pi}{2}$$

The value of $\cos \Pi(x)$ is positive, if

$$\frac{\pi}{2} > \varphi > \Pi(a)$$

and the line x is also positive.

But if $\frac{\pi}{2} > \pi - \varphi > \Pi(a)$, then the value of $\cos \Pi(x)$ is negative and the line x is on the other side of of perpendicular a.

This fact proves, that if two straight lines, lying in one plane, do not meet, regardless of how far they are extended, and are not being also parallel to each other, then both of them are perpendicular to the same straight line. Any two straight lines, which are situated in one plane and are neither parallel, nor perpendicular to one straight line must intersect.

15. Straight lines in a plane intersect mutually on the sufficient extension, if the angle between the straight line and a perpendicular from an arbitrary point on it, drawn to the other straight line, is smaller then the angle of parallelism, which corresponds the length of the perpendicular. With the help of the last suggestion we can significantly simplify the general equation of the straight line (23) in the case, when the straight line to which refers the equation, does not intersect the axis x.

Let a be a perpendicular, drawn to the axis x from a constant point, but arbitrary on a given straight line. Let L is the angle between the straight line and the perpendicular, which lies on the side x's positive (Fig. 26).

First we define the line l in such a way that:

$$\cos \Pi(l) = \tan \Pi(a) \cot L. \tag{26a}$$

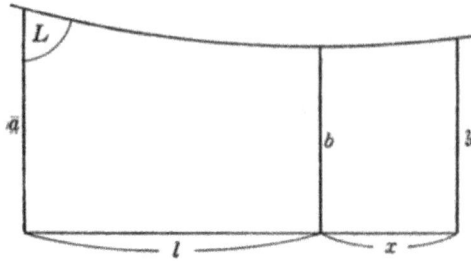

Fig. 26

This is always possible while $L > \Pi(a)$, i.e. as long as the straight line does not cross axis of x's. We put the line l on the axis x from the origin (of coordinates) on the side of x positive or negative, regardless the sign of l. Inserting at the end of the line l perpendicular to the axis x, and continue to draw it to the crossing a given straight line, and let b be a part of this perpendicular between the given straight line and the axis x. The angle under which that perpendicular meets the straight right line should be right, according to the equation (26). If we accept the end of the perpendicular b for the origin of the coordinates, then we will have according to the equation (25):

$$\cos \Pi(b) = \cos \Pi(y) \sin \Pi(x), \tag{27}$$

which will be a general equation of a straight line. which does not intersect axis x.

In this equation it can be taken $y = a$ and together with it $x = -l$; this gives:

$$\cos \Pi(b) = \cos \Pi(a) \sin \Pi(l);$$

if we put instead of $\cos \Pi, \sin \Pi(l)$ their values, then the equation becomes:

$$\cos \Pi(y) \sin \Pi(x) = \cos \Pi(a)\sqrt{1 - \tan^2 \Pi(a) \cot^2 L}.$$

The second part in the equation becomes imaginary, as soon as $\tan \Pi(a) \cot L > 1$, i.e. for each straight line, which intersects axis x.

On the basis of what we found up to this point, we can solve the problem: to define *the distance of two points* whose position in the plane was given with the help of their rectangular coordinates: x, y and x', y'. Let us put it for brevity

$$\Delta x = x' - x; \quad \Delta y = y' - y.$$

Draw from the top y (Fig. 27) a perpendicular to y' and denote the length of this perpendicular q, because y_1 denote part y' between axis x and perpendicular q.

According to equation (25), we will have:

$$\cos \Pi(y_1) = \cos \Pi(y) \sin \Pi(\Delta x),$$

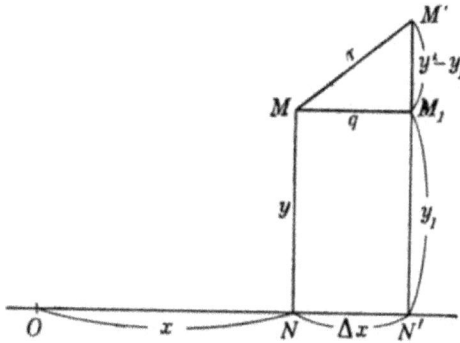

Fig. 27

$$\cos \Pi(\Delta x) = \cos \Pi(q) \sin \Pi(y_1).$$

With the help of these equations define the values of y_1 and q, the sought distance of two points, which we denote by r will be given in the following equation, which is derived from equation (11),

$$\sin \Pi(r) = \sin \Pi(y' - y_1) \sin \Pi(q).$$

8 CALCULATION OF THE LENGTH OF THE ARC OF A PLANE CURVED LINE

16. If Δx and Δy and therefore q, r are extremely small, so we can neglect their high orders of magnitude compared to the lower orders, then r will present the element ds of a curved line, to which expression we arrive by taking

$$\sin \Pi(q) = 1 - \frac{1}{2}q^2,$$

$$\cos \Pi(q) = q - \frac{1}{3}q^3,$$

$$\sin \Pi(r) = 1 - \frac{1}{2}r^2,$$

$$\sin \Pi(y' - y_1) = 1 - \frac{1}{2}(y' - y_1)^2,$$

after that we get

$$q = \frac{\Delta x}{\sin \Pi(y)},$$

$$ds = \sqrt{dy^2 + \frac{dx^2}{\sin^2 \Pi(y)}}. \qquad \text{[A]}$$

For the limiting circle:

$$\sin \Pi(y) = e^{-x}.$$

From the general expressions, which define $\sin \Pi(a)$ and so on with the help of a and which are given above we have:

$$d\Pi(a) = -\sin \Pi(a)da, \qquad \text{[I]}$$

after that we differentiate the equation of the limiting circle and find:

$$\sin \Pi(y) \cos \Pi(y)dy = e^{-x}dx$$

and

$$ds = \frac{dxe^x}{\sqrt{1 - e^{-2x}}}. \qquad \text{[A']}$$

157

158

Integrate with respect to x from $x = 0$, we find:

$$s = \sqrt{e^{2x} - 1}$$

or otherwise

$$s = \cot \Pi(y)$$

as it was found above.

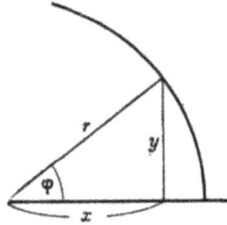

Fig. 28

If we denote by r the distance to a point on the circular line from the origin of the coordinates, and φ is the angle which this distance r makes with the axis x (positive values of x) (Fig. 28), then we find in the right triangle, whose sides are y, x, r according to equation (12):

$$\tan \Pi(r) = \sin \varphi \tan \Pi(y).$$

Taking logarithms from both sides of this equation and differentiating it with respect to r, φ, y we get:

$$\frac{dr}{\cos \Pi(r)} = -\cot \varphi d\varphi + \frac{dy}{\cos \Pi(y)}.$$

From this equation above we derive:

$$dy = \left\{ \cot \varphi d\varphi + \frac{dr}{\cos \Pi(r)} \right\} \cos \Pi(y), \tag{27a}$$

or use, instead of $\cos \Pi(y)$ its value with the help of φ and r we have:

$$dy = \frac{\cos \varphi \cos \Pi(r) d\varphi + \sin \varphi dr}{\sqrt{1 - \cos^2 \varphi \cos^2 \Pi(r)}}.$$

In order to express dx with the help of r, φ, we take equation (10)

$$\cos \Pi(r) \cos \varphi = \cos \Pi(x).$$

Differentiating logarithms of both sides of this equation with respect to r, φ, x we get:

$$\frac{\sin^2 \Pi(r) dr}{\cos \Pi(r)} - \tan \varphi d\varphi = \frac{\sin^2 \Pi(x) dx}{\cos \Pi(x)}, \tag{27b}$$

from where we derive with the help of the following equation:

$$\sin \Pi(x) \sin \Pi(y) = \sin \Pi(r),$$

$$\cos \Pi(r) \cos \varphi = \cos \Pi(x),$$

the following equation, which expresses the sought value dx is:

$$\frac{dx}{\sin \Pi(y)} = \frac{\cos \varphi \sin \Pi(r) dr - \sin \varphi \cot \Pi(r) d\varphi}{\sqrt{1 - \cos^2 \varphi \cos^2 \Pi(r)}},$$

after that (Fig. 29)

$$ds = \sqrt{dr^2 + d\varphi^2 \cot^2 \Pi(r)}. \qquad [B]$$

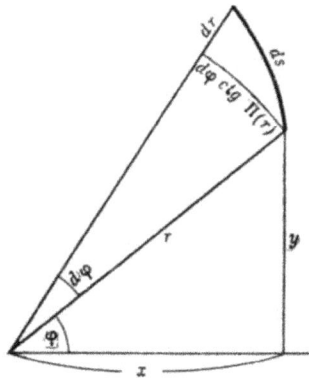

Fig. 29

For the circle, assuming that the origin of the coordinate is in the center of the circle we find as long as we have here $dr = 0$ the following equation:

$$ds = d\varphi \cot \Pi(r).$$

Integrating from $\varphi = 0$ to $\varphi = \frac{\pi}{2}$ and multiplying everything by 4, we find the following expression for circumference with the radius r:

$$2\pi \cot \Pi(r),$$

according to what was found above.

If we call s the arc of the limiting circle from the axis x, then a rotation s around the axis x creates part of the limiting sphere, and the end of the arc will draw a circumference of the circle, which on the limiting sphere is defined the same way as the circumference of the circle with a radius s defines on the planes in the ordinary geometry. From here it follows, that circumference of this circle should be equal to $2\pi s$. On the other hand the circumference of the same circle, considered in its plane, where the perpendicular y drawn from the end of the arc s to the axis of the limiting circle, which serves axis x and passes through the other end of the arc, makes, constitutes a radius of the circle, is given in pangeometry by the expression:

$$2\pi \cot \Pi(y);$$

from where it follows, that $s = \cot \Pi(y)$, as it was proven before.

9 CALCULATING THE AREA OF PLANE FIGURES

17. In order that we divide the area into elements, we will introduce on the area arcs of the limiting circle, whose axis will be axis x and in such a way, that their mutual distance be infinitesimal and can be called ds (Fig. 30). Let s be one of these arcs $[MN]$ of the limiting circle between axe x and a point on the curved line, whose coordinates are x, y; let s' be an arc from another of these limiting circles $[M'N']$ between axis and point $[N']$ on this curved line, whose coordinates are $x + dx$, $y + dy$.

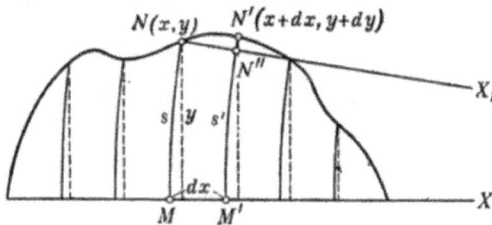

Fig. 30

This part of the area, which is situated between s and s' with one side and the axe x and the curved from the other is infinitesimally small and is expressed by the equation:

$$dS = \frac{es\,dx}{e-1},$$ [a]

in the above equation we insert $s = \cot \Pi(y)$ we get:

$$dS = \frac{edx \cot \Pi(y)}{e-1}.$$ [b]

For an example we define the area of the limiting circle (Fig. 31), for which we have found an equation in the rectangular coordinates:

$$\sin \Pi(y) = e^{-x},$$

with the help of the above equation we will find equation for element of the sought area:

$$dS = \frac{e}{e-1}dy \cos \Pi(y) \cot \Pi(y).$$

Integrating this expression from $y = 0$, we find the area $[ABC]$, bounded

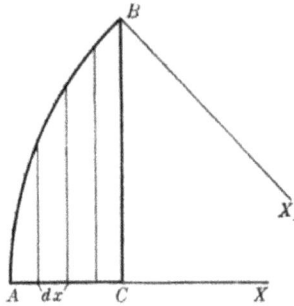

Fig. 31

by the arc of the limiting circle axis x and ordinate y

$$S = \frac{e}{e-1}\left\{ \cot \Pi(y) - \frac{1}{2}\pi + \Pi(y) \right\}.$$

We saw, that the area between two parallel lines, extended to infinity on the side of the parallelism and bounded by the arc of the limiting circle, whose arc have two parallels for axes, is expressed in such way:

$$\frac{es}{e-1} = \frac{e \cot \Pi(y)}{e-1},$$

after that we find the area between two straight parallel lines, from which one is perpendicular to y, which is drawn through the ends of y and extended to infinity, is expressed

$$\frac{1}{2}\pi - \Pi(y). \tag{27c}$$

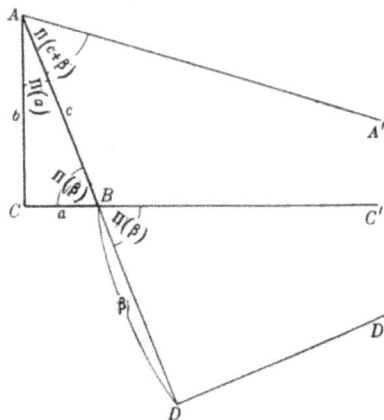

Fig. 32

With the help of the above expression we find the area of rectilinear right triangle, when the given astute angles $\Pi(\alpha), \Pi(\beta)$ which are opposite to the perpendiculars a, b of the triangle (Fig. 32). We extend the hypotenuse c beyond the top $[B]$ of the angle $\Pi(\beta)$ and we make the extension $[BD]$ equal to β. The perpendicular $[DD']$ at the end of β is made parallel to the extension of the side a. The area $[D'DBC']$, infinitely stretched between these two parallels, extended in the side of the parallelism and bounded on the other side by the line β, will be

$$\frac{1}{2}\pi - \Pi(\beta).$$

If we now draw through the apex A the angle $\Pi(\alpha)$ a straight line parallel to the perpendicular DD' and which is inclined to the side c under an angle $\Pi(c+\beta)$ and will be also parallel to the extension of the side a, then the magnitude of the area $D'DAA'$ between $c+\beta\,[AD]$ and two parallels through the ends $c+\beta$, drawn and extended to infinity in the side of parallelism, will be:

$$\frac{1}{2}\pi - \Pi(c+\beta).$$

In a similar way, part of the area $[A'ACC']$ between the side b, the straight line AA', drawn through the top $\Pi(\alpha)$, and the side $a\,[CB]$, infinitely extended, will be

$$\frac{1}{2}\pi - \Pi(b).$$

After that the sum $\frac{1}{2} - \Pi(\beta)$, $\frac{1}{2}\pi - \Pi(b)$ decreased [by] $\frac{1}{2}\pi - \Pi(c+\beta)$, will be the expression of the area of the triangle $[ABC]$, which in such a way find $= \frac{1}{2}\pi - \Pi(b) - \Pi(\beta) + \Pi(c+\beta)$. Meanwhile we proved, that $\Pi(b) = \Pi(\alpha) + \Pi(c+\beta)$; placing from here into the expression of the area of triangle on the place of $\Pi(b)$ its value, we find the area of rectilinear and right triangle:

$$\frac{1}{2}\pi - \Pi(\alpha) - \Pi(\beta);$$

this means, that the area of rectilinear and right triangle is equal to the difference of the two right angles without the sum of the three angles of the triangle. From here it follows as well that the area of each rectilinear triangle is equal to the surplus of the two right angles over the sum of the three angles of the triangle.

It is easy to derive from the previous that the area of each quadrangle is equal to the surplus of the four right angles over the sum of the four angles of the quadrangle and in general *the area of the poligon n side is equal to the surplus $(n-2)\pi$ over the sum of the angles of the poligon.*

18. Considering especially quadrangle, which two sides a, y are both perpendiculars to third side x and which fourth side t is perpendicular to the side a and does with y angle which will call ω (Fig. 33).

We proved above (equation 25), that between components of such quadrangle there exists an equation

$$\cos\Pi(\alpha) = \cos\Pi(y)\sin\Pi(x);$$

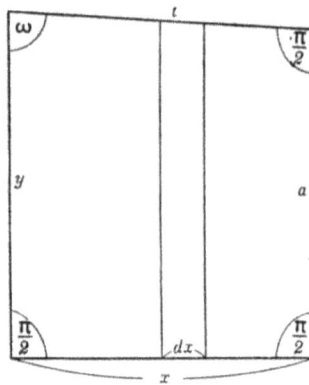

Fig. 33

if we consider x and y as variables and a constant, then the area of the quadrangle is expressed, like any area, as it is proven above by the integral

$$\int dx \cot \Pi(y) \qquad (27\text{d})$$

which, being applied to the present case, gives when put here value $\cot \Pi(y)$ in the following equation for area:

$$\int_0 \frac{dx \cos \Pi(a)}{\sqrt{\sin^2 \Pi(x) - \cos^2 \Pi(a)}},$$

then as the area of this quadrangle, according to how we define the area of each polygon with the help of angles $= \frac{1}{2}\pi - \omega$. This gives:

$$\frac{1}{2}\pi - \omega = \cos \Pi(a) \int_0 \frac{dx}{\sqrt{sin^2(a) - \cos^2 \Pi(x) - \cos^2 \Pi(a)}}. \qquad (\text{M})$$

The angle ω between the sides t and y, is defined by the equation (equation 26)

$$\tan \omega = \frac{\tan \Pi(y)}{\cos \Pi(x)}.$$

If in the equation (M) we write α instead of $\Pi(a)$, ξ instead of $\Pi(x)$, then we will have

$$\frac{\frac{1}{2}\pi - \omega}{\cos \alpha} = \int_{\frac{\pi}{2}} \frac{d\xi}{\sin \xi \sqrt{sin^2 \alpha - \cos^2 \xi}},$$

where α is a constant quantity.

We can check this equation for the integral through differentiation. Pangeometria also points out a new method to the values of given integrals.

The area between two parallels, drawn from the ends of the given straight line, extended to infinity in the side of parallels and the straight

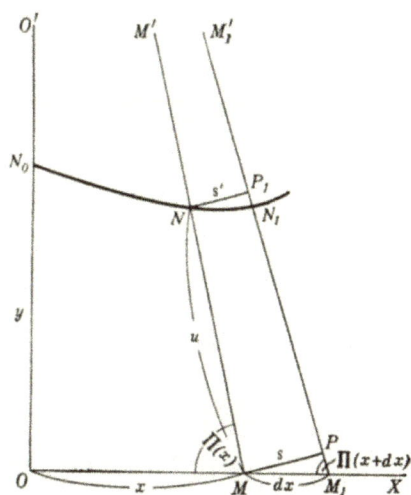

Fig. 34

line will be $= \pi - [$ minus$]$ the sum of the two angles, which two parallel make with a given straight line, because this area could be taken for the area of a triangle, where one angle is equal to zero.

The area of a curved line could be divided into elements of straight parallel lines to the same given straight line, for example, to the axis y (Fig. 34). If we draw the axis x parallel to the line $[MM']$ to the axis y then this straight line forms with the axis x the angle $= \Pi(x)$; the straight line, drawn through the end of the axis $x + dx$, makes in such a way with the axis x the angle $= \Pi(x + dx)$, from where it follows, that the area between these two parallels and dx is equal to $-d\Pi(x)$. Let now u be a quantity of the first parallel between axis x and the curved; this part of the area between the two parallels, which lies out of the given curve, will be a consequence of what was proved above

$$-e^{-u}d\Pi(x),$$

from where it follows, that the part of the area, which lies between the curve and the axis x, i.e. the element of the area of the curve, is expressed by

$$.dS = -(1 - e^{-u})d\Pi(x) \tag{27e}$$

In order to calculate the area of a circle with a radius r, it is necessary to put the value of $\cot \Pi(y)$ in the general equation of the area element of the curve, found above,

$$dS = dx \cot \Pi(y),$$

insert the expression $\cot \Pi(y)$ from the equation of the circle:

$$\sin \Pi(x) \sin \Pi(y) = \sin \Pi(r),$$

where the origin of the rectilinear coordinates is in the center of the circle. This gives the following equation

$$dS = dx\sqrt{\frac{\sin^2 \Pi(x)}{\sin^2 \Pi(r)} - 1};$$

integrating $x = 0$, we find:

$$S = \frac{1}{\sin \Pi(r)} \arcsin\left\{\frac{\cos \Pi(x)}{\cos \Pi(r)}\right\} - \arcsin\left\{\frac{\cot \Pi(x)}{\cot \Pi(r)}\right\}.$$

For $x = r$ this gives the area of a quarter of the circle:

$$\frac{\pi}{2\sin \Pi(r)} - \frac{\pi}{2}.$$

Multiplying by 4, we find for the area of the whole circle

$$2\pi\left\{\frac{1}{\sin \Pi(r)} - 1\right\}$$

or which is equal to:

$$\pi\{e^{\frac{r}{2}} - e^{-\frac{r}{2}}\}^2.$$

If r is very small, then this expression gives the area of the circle $= \pi r^2$; the same expression, which is given usually for the area of the circle in Geometry.

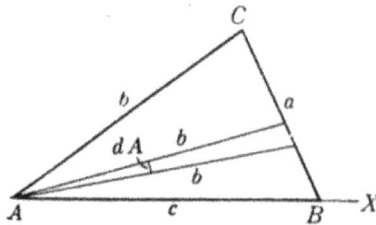

Fig. 35

19. With the help of the previous expression of the area of the circle we can give the area element of any curved line also the following expression:

$$dS = d\varphi\left\{\frac{1}{\sin \Pi(r)} - 1\right\}, \tag{27f}$$

where r is a radius vector, drawn from the origin of the coordinates, to a point on the curved line, and φ angle, which this radius vector makes with the constant straight line, passing through the origin of the coordinate [Fig. 29].

The application of this equation to the calculation the area of the triangle (Fig. 35), whose sides a, b, c with opposite angles A, B, C gives, if we consider A, C, and the sides a, b as variables:

$$\text{the area of the triangle} \quad = \int_0^A dA\left\{\frac{1}{\sin \Pi(b)} - 1\right\}.$$

The side b is expressed as a function of c, A, B by the help of the last equation (19)

$$\cot B \sin A \sin \Pi(c) + \cos A = \frac{\cos \Pi(c)}{\cos \Pi(b)}.$$

We take from this equation the value $\sin \Pi(b)$ and put it in the expression for the area of the triangle, we get:

$$\text{the area of the triangle} \quad = \int_0^A \frac{dA}{\sqrt{1 - \dfrac{\cos^2 \Pi(c)}{(\cot B \sin A \sin \Pi(c) + \cos A)^2}}} - A.$$

Meanwhile it was proven, that the area of the triangle

$$= \pi - A - B - C,$$

where the angles A and B are given and C is defined by equation (19)

$$\cos C + \cos A \cos B = \frac{\sin A \sin B}{\sin \Pi(c)}.$$

Comparing these two expression for the area of the triangle gives:

$$\pi - B - C = \int_0^A \frac{dA\{\cot B \sin A \sin \Pi(c) + \cos A\}}{\sqrt{(\cot B \sin A \sin \Pi(c) + \cos A)^2 - \cos^2 \Pi(c)}}.$$

If $B = \frac{\pi}{2}$, this equation gives:

$$\frac{\pi}{2} - C = \int_0^A \frac{dA \cos A}{\sqrt{\cos^2 A - \cos^2 \Pi(c)}},$$

an equation, which after integration becomes:

$$\frac{1}{2}\pi - C = \arcsin\left(\frac{\sin A}{\sin \Pi(c)}\right),$$

which is in accordance with the equation, which defines C.

Based on what we proved above, we can derive for the expression of the area of each closed polygon two expressions, one of them given by the integral, the other depending only on the sum of the angles of the polygon.

The two values for the same area, being necessarily equal, give a method to calculate definite integrals, whose value is difficult to find in another way.

In order to give more examples, we consider rectilinear right triangle, whose perpendicular sides are x, y, a hypotenuse is r (Fig. 36). Let the angle A be opposite to y, and angle B be opposite to x. Equations (10), (11) are give for such a triangle:

$$\sin \Pi(x) \sin \Pi(y) = \sin \Pi(r),$$
$$\sin \Pi(x) \cos B = \sin A,$$
$$\cos \Pi(r) \cos A = \cos \Pi(x),$$
$$\cos \Pi(r) \cos B = \cos \Pi(y).$$

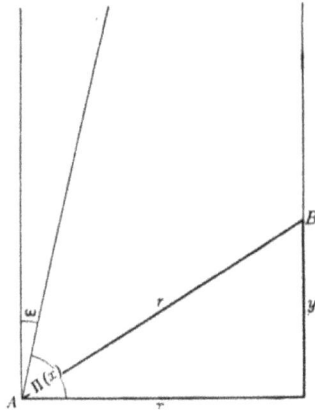

Fig. 36

From the these equations we derive:

$$\cos \Pi(r) = \frac{\cos \Pi(x)}{\cos A},$$

$$\sin \Pi(r) = \sqrt{1 - \left(\frac{\cos \Pi(x)}{\cos A}\right)^2},$$

$$\sin \Pi(y) = \frac{1}{\sin \Pi(x)} \sqrt{1 - \left(\frac{\cos \Pi(x)}{\cos A}\right)^2} =$$

$$= \sqrt{\frac{1}{\sin^2 \Pi(x)} - \frac{\cot^2 \Pi(x)}{\cos^2 A}},$$

$$\cot \Pi(y) = \frac{\sin A \cos \Pi(x)}{\sqrt{\cos^2 A - \cos^2 \Pi(x)}}.$$

Inserting in the last equation $\Pi(x) = \frac{\pi}{2} - \omega$, we find:

$$\cot \Pi(y) = \frac{\sin A \sin \omega}{\sqrt{\cos^2 A - \sin^2 \omega}}.$$

Meanwhile we saw, that the differential of the area is $dx \cot \Pi(y)$; and this gives in the present case:

$$dx \cot \Pi(y) = \sin A \frac{d\omega \tan \omega}{\sqrt{\cos^2 A - \sin^2 \omega}};$$

from where we conclude, integrating from $\omega = 0$, which corresponds $x = 0$ and noticing that the area expressed by integral is expressed also through $\frac{\pi}{2} - A - B$, that

$$\frac{\pi}{2} - A - B = \sin A \int^\omega \frac{\tan \omega \, d\omega}{\sqrt{\cos^2 A - \sin^2 \omega}},$$

where angle A is constant, and B is defined by the equation

$$\cos B = \frac{\sin A}{\cos \omega}.$$

If $\omega = \frac{\pi}{2} - A$, the then hypotenuse is parallel to the side y and angle $B = 0$. So in this case

$$\frac{\pi}{2} - A = \sin A \int_0^{\frac{\pi}{2}-A} \frac{\tan \omega \, d\omega}{\sqrt{\cos^2 A - \sin^2 \omega}}.$$

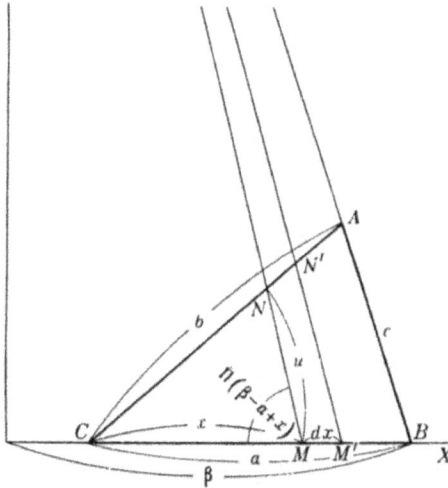

Fig. 37

It is possible to determine the value of the integral in a more general form, considering the area of the rectilinear triangle, with sides a, b, c with opposite angles A, B, C and dividing this area into elements by straight lines and parallel between each other. Take the top of the angle C (Fig. 37) for the origin of the coordinates and the side a for the axis x. Let $B = \Pi(\beta)$, where β is positive if $B < \frac{\pi}{2}$ and negative if $B > \frac{\pi}{2}$. Draw through the end $[M]$ of the axis x the straight line u $[MN]$ parallel to side c, and continue to the intersection with the side b. The angle, which this parallel makes with the axis x, will be $\Pi(\beta - a + x)$ from where it follows that the angle which makes this parallel with the extension of x, will be $\Pi(a - \beta - x)$. If we take for the element of the area of the triangle part of the area, which is between two parallel u infinitesimally close then we will get, based on what was proved above, the following equation for that element:

$$dS = -d\Pi(a - \beta - x)\{1 - e^{-u}\}.$$

Consider y, x, u as variable and a and β as constant.

The equations (19), being applied to the triangle, whose sides are x, u and the angle between the two sides $\Pi(\beta - a + x)$, give:

$$\cot C \sin \Pi(\beta - a + x) \sin \Pi(x) + \cos \Pi(\beta - a + x) = \frac{\cos \Pi(x)}{\cos \Pi(u)};$$

from this equation we derive, putting for brevity $\Pi(\beta - a + x) = \omega$,

$$\cos \Pi(u) = \frac{\cos \Pi(x)}{\cot C \sin \omega \sin \Pi(x) + \cos \omega},$$

$$e^{2u} = \frac{\cot C \sin \omega \sin \Pi(x) + \cos \omega + \cos \Pi(x)}{\cot C \sin \omega \sin \Pi(x) + \cos \omega - \cos \Pi(x)};$$

but

$$\sin \Pi(x) = \sin \Pi\{(\beta - a) - (\beta - a + x)\} = \frac{\sin \Pi(\beta - a) \sin \omega}{1 - \cos \Pi(\beta - a) \cos \omega}.$$

In such a way we find:

$$\cos \Pi(x) = \frac{\cos \Pi(\beta - a) - \cos \omega}{1 - \cos \Pi(\beta - a) \cos \omega}.$$

Inserting the values $\sin \Pi(x)$, $\cos \Pi(x)$ in the equation for e^{2u} we get:

$$e^{2u} =$$

$$\frac{\cot C \sin^2 \omega \sin \Pi(\beta - a) + \cos \omega\{1 - \cos \Pi(\beta - a) \cos \omega\} + \cos \Pi(\beta - a) - \cos \omega}{\cot C \sin^2 \omega \sin \Pi(\beta - a) + \cos \omega\{1 - \cos \Pi(\beta - a) \cos \omega\} - \cos \Pi(\beta - a) + \cos \omega}$$

$$= \frac{\cot C \sin^2 \omega \sin \Pi(\beta - a) + \cos \Pi(\beta - a) \sin^2 \omega}{\cot C \sin^2 \omega \sin \Pi(\beta - a) + 2 \cos \omega - \{1 + \cos^2 \omega\} \cos \Pi(\beta - a)};$$

further we find:

$$d\Pi(a - \beta - x) = -d\Pi(\beta - a + x) = -d\omega,$$

after that the above comparison of two expressions for the area of the triangle we get the equation:

$$\pi - A - B - C = -\omega + \int_{x=0}^{x=a} d\omega \times$$

$$\times \sqrt{\frac{\cot C \sin^2 \omega \sin \Pi(\beta - a) + 2 \cos \omega - (1 + \cos^2 \omega) \cos \Pi(\beta - a)}{\cot C \sin^2 \omega \sin \Pi(\beta - a) + \cos \Pi(\beta - a) \sin^2 \omega}}.$$

If we also put $\Pi(\beta - a) = \alpha$, then this equation will have the following form:

$$[\pi - A - B - C + \alpha - \Pi(\beta)][\cot C \sin \alpha + \cos \alpha]^{\frac{1}{2}} =$$

$$= \int_{\omega=\alpha}^{\omega=\Pi(\beta)} \frac{d\omega}{\sin \omega} \sqrt{\cot C \sin^2 \omega \sin \alpha + 2 \cos \omega - (1 + \cos^2 \omega) \cos \alpha},$$

where the angles A, B and the line β should be calculated by the equation:

$$\alpha - \Pi(\beta - a); \quad B = \Pi(\beta),$$

$$\cos A + \cos B \cos C = \frac{\sin B \sin C}{\sin \Pi(a)}.$$

The last of these equations is the last of equations (19), applied to the triangle, which we considered.

10 LIMITTING COORDINATES

20. In order to define the position of a point in the plane, we can use in pangeometry not only rectilinear and polar coordinates, but also the arcs of the limiting circle. Even the last system is very useful for the simplicity of the expression. Define the position of point $[M]$ (Fig. 38) in the plane with the help of perpendicular coordinates x and y in such a way that y is a perpendicular from a point, which position we like to define, on the axis x, and x is a part of the axis x of the perpendicular $[N]$ to the origin of the coordinates. Let η be the length of the arc $[PM]$ of the limiting circle from a given point to the axis x, which together serve as the axis of the limiting circle; call ξ the distance $[OP]$ of the top of the limiting circle on the axis x to the origin of the coordinates. We saw that in this case

$$\eta = \cot \Pi(y),$$

after the equation of the limiting circle gives

$$e^{-(x-\xi)} = \sin \Pi(y);$$

with the help of these two equations we can express ξ, η depending on x, y or vice versa x, y depending on $\xi \eta$. This allows to go from the equation of a curved line into coordinates x, y to the equation of the same curved line into ξ, η or vice versa.

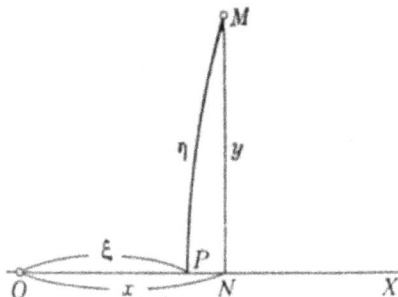

Fig. 38

The differential of the area is expressed in $\xi \eta$ equations (Fig. 39)

$$d^2 S = d\xi d\eta,$$

171

where S is the area.

If we consider S as a function x, y then we have:

$$\left(\frac{dS}{dx}\right) = \frac{dS}{d\xi},$$

and after differentiating with respect to y:

$$\frac{d^2 S}{dxdy} = \frac{1}{\sin \Pi(y)} \frac{d^2 S}{d\xi d\eta} = \frac{1}{\sin \Pi(y)}$$

according to what was found above.

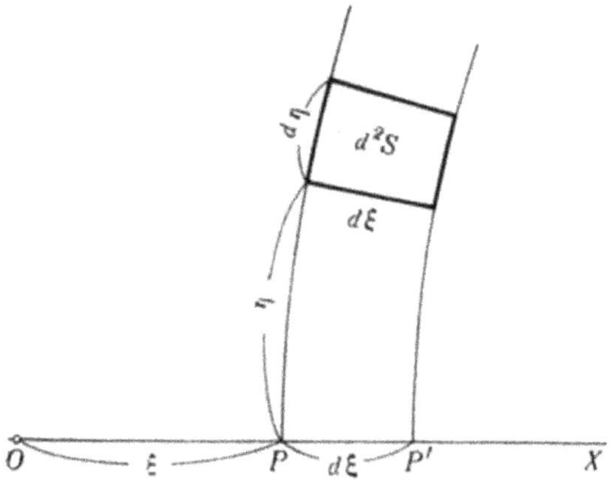

Fig. 39

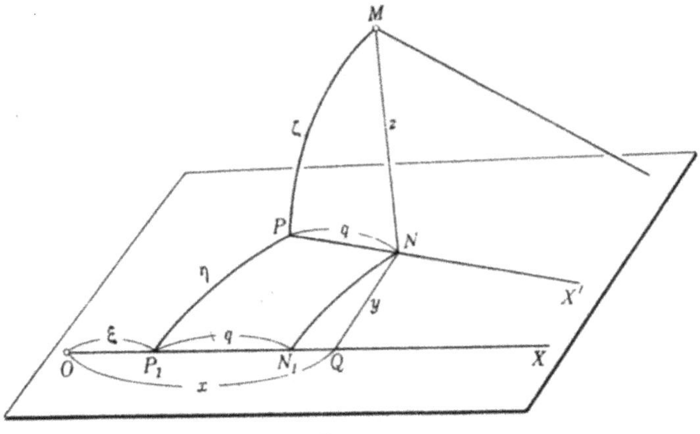

Fig. 40

Draw from point $[M]$ in the space the perpendicular z, $[MN]$ on the plane of coordinates x, y (Fig. 40).

Draw through this perpendicular a plane, which intersect the plane xy in the straight line which is parallel to the axis x. Consider this intersection $[NX']$ on the side of the parallelism for the axis of the limiting circle, which goes through the apex $[M]$ perpendicular to z, and let ζ is the length of the arc $[MP]$ of this limiting circle between the apex z and the axis. Then we get:

$$\zeta = \cot \Pi(z).$$

The part q parallel to the axis x, drawn through the end of the perpendicular z between the top $\zeta[P]$ and the end $[N]$ of the perpendicular z be given by the equation

$$e^{-q} = \sin \Pi(z).$$

The arc $[NN_1]$ of the limiting circle, drawn through the end z in the way that the axis of positive x serves instead of axis of limiting circle and which is situated between the end of z and this axis, will be equal to $\cot \Pi(y)$, and the length of the arc $\eta[PP_1]$ of the limiting circle, drawn through the intersection ζ with the plane x, y and for which axis lies on side of the positive x, between this point and the axis will be, as it was proven, given by the following equation:

$$\eta = \frac{\cot \Pi(y)}{\sin \Pi(z)}.$$

If we call ξ part of the axis $x's$ between the origin of the coordinates and the arc η, then the equation of the limiting circle gives:

$$e^{-x+\xi+q} == \sin \Pi(y).$$

From these equations we derive, changing in advance only z and depending on it ζ:

$$d\zeta = \frac{dz}{\sin \Pi(z)}.$$

Changing only y and η, we get:

$$d\eta = \frac{dy}{\sin \Pi(y) \sin \Pi(z)}.$$

At the end changing only ξ and x, we get:

$$d\xi = dx.$$

11 EXPRESSING THE AREA OF A TRIANGLE THROUGH ITS SIDES

21. In order to complement the new theory called pangeometry, which is based on the more general foundations rather than that of the ordinary geometry, it remains only to give an expression for the differential of the surface and the volume with the help of coordinates, defining the position of a point in space.

Considering to this end again a quadrangle where two sides a, y are perpendicular to the third x and where the fourth side c is perpendicular to y, which makes with a an angle φ. We have found (equation 25):

$$\cos \Pi(y) = \cos \Pi(a) \sin \Pi(x).$$

After that we find with the help of the equations (10), (11), calling r diagonal between the top of the angle φ and the top of the right angle which is opposite and A the angle between x and r:

$$\cos \Pi(r) \cos A = \cos \Pi(x),$$
$$\cos A \tan \Pi(c) = \tan \Pi(r).$$

From these two equations we draw:

$$\cos \Pi(x) \tan \Pi(c) = \sin \Pi(r);$$

but

$$\sin \Pi(r) = \sin \Pi(a) \sin \Pi(x),$$

hence:

$$\tan \Pi(c) = \sin \Pi(a) \tan \Pi(x).$$

If c, x are so small, that we can neglect their high powers in comparison to the low powers and allow the following approximate values for $\tan \Pi(c), \tan \Pi(x)$:

$$\tan \Pi(c) = \frac{1}{c}; \quad \tan \Pi(x) = \frac{1}{x},$$

then we find

$$c = \frac{x}{\sin \Pi(a)} \tag{27'}$$

175

The straight lin *ce* which connects the top a, y, will not be perpendicular to y if $a = y$ in the quadrangle; in such a case the straight line p from the middle of x drawn to the middle of c, is perpendicular to x and to c. So we can replace in the equation (27') c by $\frac{c}{2}$, and x can be replaced by $\frac{x}{2}$ from which the equation does not change. In such a way this equation can be proved even for the following case $a = y$ where the above given proof is not applicable directly.

The quantity of the curved surface is measured by the sum of the areas of the triangles, which blend into a continuous set and the top, which lies on the surface. This measure will be as accurate as the measured triangles are smaller.

The limit, to which this sum approaches infinity, when the measured triangles are getting infinitesimally and this quantity is called *mathematical quantity of the surface.*

First define the area of the rectilinear right triangle with the help of its three sides abc with the opposite angles $\Pi(\alpha), \Pi(\beta), \frac{\pi}{2}$. We saw that, that in this triangle one can replace the lines

$$a, b, c, \alpha, \beta$$

by the lines:

$$a, \alpha', \beta, b', c$$

respectively.

Except this we found:

$$2\Pi(b) = \Pi(c + \beta) + \Pi(c - \beta).$$

Inserting α' instead of b, β instead of c, and c instead of β, we get

$$\pi - 2\Pi(\alpha) = \Pi(\beta + c) + \Pi(\beta - c)$$

or

$$2\Pi(\alpha) = \Pi(c - \beta) - \Pi(c + \beta).$$

In this way we find:

$$2\Pi(\beta) = \Pi(c - \alpha) - \Pi(c + \alpha).$$

Changing in this equation the letters, as it was said, we get:

$$2\Pi(c) = \Pi(\beta - b') - \Pi(\beta + b').$$

In such a way we find:

$$2\Pi(c) = \Pi(\alpha - a') - \Pi(\alpha + a'),$$

from where with the changed letters, as it was indicated above, we derive:

$$2\Pi(\beta) = \Pi(b' - a') - \Pi(b' + a').$$

In such a way we get:

$$2\Pi(\alpha) = \Pi(a' - b') - \Pi(a' + b').$$

Adding the two last equations gives:

$$2\Pi(\alpha) + 2\Pi(\beta) = \pi - 2\Pi(a' + b').$$

After that the area of the triangle $\triangle$ is given by the expression:

$$\triangle = \frac{\pi}{2} - \Pi(\alpha) - \Pi(\beta) = \Pi(\beta) = \Pi(a' + b')[331]$$

and then:

$$\tan\frac{1}{2}\triangle = e^{-a'}e^{-b'} = \tan\left\{\frac{1}{4}\pi - \frac{1}{2}\Pi(a)\right\}\tan\left\{\frac{1}{4}\pi - \frac{1}{2}\Pi(b)\right\};$$

from where at the end we derive:

$$\tan\frac{1}{2}\triangle = \frac{e^a - 1}{e^a + 1}\frac{e^b - 1}{e^b + 1}.$$

When a, b are so small that we can neglect the higher powers from $a, b, \triangle$, then these formula gives: $\triangle = \frac{1}{2}ab$, as in the ordinary geometry.

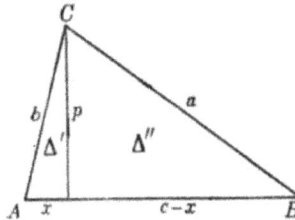

Fig. 41

In the rectilinear triangle we can always choose one angle C in such a way, that from its top a perpendicular to the opposite side lies inside the triangle (Fig. 41). This perpendicular will divide side c of the triangle in two parts: one is x, adjacent to the angle A, the other $c - x$, adjacent to the angle B. The area S of this triangle will be equal to the sum of the areas of the two right triangles, formed by this perpendicular h and will be given by the equation:

$$\tan\frac{1}{2}S = \frac{\frac{e^x - 1}{e^x + 1}\frac{e^h - 1}{e^h + 1} + \frac{e^{c-x} - 1}{e^{c-x} + 1}\frac{e^h - 1}{e^h + 1}}{1 - \frac{e^x - 1}{e^x + 1}\frac{e^{c-x} - 1}{e^{c-x} + 1}\left(\frac{e^h - 1}{e^h + 1}\right)}.$$

The equation which could be given the following form:

$$\tan\frac{1}{2}S = \frac{(e^{2h} - 1)(e^c - 1)}{(e^x + e^{c-x})(e^h + 1)^2 + 2e^h(e^x - 1)(e^{c-x} - 1)}.$$

This expression gives, if we neglect the higher powers S, h, c before the lower powers:

$$S = \frac{1}{2}ch,$$

as in the ordinary geometry.

22. We saw, that the area of a triangle is expressed by three angles A, B, C of the triangle in the following way:

$$S = \pi - A - B - C.$$

Take value of A depending of a, b, c from the second equation (19); and we get:

$$\cos A = \frac{1 - \frac{\sin \Pi(b) \sin \Pi(c)}{\sin \Pi(a)}}{\cos \Pi(b) \cos \Pi(c)};$$

from where it follows

$$2 \cos^2 \frac{1}{2} A = \frac{1 + \cos \Pi(b) \cos \Pi(c) - \frac{\sin \Pi(b) \sin \Pi(c)}{\sin \Pi(a)}}{\cos \Pi(b) \cos \Pi(c)}$$

If we put here:

$$\frac{\sin \Pi(b) \sin \Pi(c)}{\sin \Pi(b + c)}$$

instead of

$$1 + \cos \Pi(b) \cos \Pi(c),$$

then this expression will take the following form:

$$2 \cos^2 \frac{1}{2} A = \tan \Pi(b) \tan \Pi(c) \left\{ \frac{1}{\sin \Pi(b + c)} - \frac{1}{\sin \Pi(a)} \right\}.$$

In such a way we find:

$$-2 \sin^2 \frac{1}{2} A = \tan \Pi(b) \tan \Pi(c) \left\{ \frac{1}{\sin \Pi(b - c)} - \frac{1}{\sin \Pi(a)} \right\}.$$

From these two expressions we derive:

$$\sin^2 A = \tan^2 \Pi(b) \tan^2 \Pi(c) \times$$

$$\times \left\{ -\frac{1 - \cos^2 \Pi(b) \cos^2 \Pi(c)}{\sin^2 \Pi(b) \sin^2 \Pi(c)} + \frac{2}{\sin \Pi(a) \sin \Pi(b) \sin \Pi(c)} - \frac{1}{\sin^2 \Pi(a)} \right\}$$

or

$$\sin^2 A = -\tan^2 \Pi(b) \tan^2 \Pi(c) \times$$

$$\times \left\{ \frac{1}{\sin^2 \Pi(a)} + \frac{1}{\sin^2 \Pi(b)} + \frac{1}{\sin^2 \Pi(c)} - \frac{2}{\sin \Pi(a) \sin \Pi(b) \sin \Pi(c)} - 1 \right\}.$$

Assuming for brevity the following:

$$P = \sqrt{\frac{-1}{\sin^2 \Pi(a)} - \frac{1}{\sin^2 \Pi(b)} - \frac{1}{\sin^2 \Pi(c)} + \frac{2}{\sin \Pi(a) \sin \Pi(b) \sin \Pi(c)} + 1},$$

we get:

$$\sin A = \tan \Pi(b) \tan \Pi(c) P. \tag{28}$$

Also we can give P the following form, symmetrical with respect to a, b, c:

$$P^2 = 2\left\{1 + \frac{1}{\sin \Pi(a)}\right\}\left\{1 + \frac{1}{\sin \Pi(c)}\right\} - \left\{1 + \frac{1}{\sin \Pi(a)} + \frac{1}{\sin \Pi(b)} + \frac{1}{\sin \Pi(c)}\right\}^2.$$

Coming from the equation (28) and considering here P as undefined quantity, we can prove in the following way that P should be a symmetric function of a, b, c. Multiplying equation (28) by $\tan \Pi(a)$, we put here $\sin B \tan \Pi(b)$ instead of its value $\sin A \tan \Pi(a)$ (equation 13) and after divide by $\tan \Pi(b)$, we will have:

$$\sin B = \tan \Pi(a) \tan \Pi(c) P.$$

Multiplying this last equation by $\tan \Pi(b)$ and inserting here $\sin C \tan \Pi(c)$ instead of its value $\sin B \tan \Pi(b)$ (equation 13) and dividing after by $\tan \Pi(c)$; we get:

$$\sin C = \tan \Pi(a) \tan \Pi(b) P,$$

from where it is seen, that the function P is symmetrical with respect to a, b, c:

We have already found:

$$\cos A = \frac{1 - \frac{\sin \Pi(b) \sin \Pi(c)}{\sin \Pi(a)}}{\cos \Pi(b) \cos \Pi(c)}$$

or, it is equal to:

$$\cos A = \tan \Pi(b) \tan \Pi(c)\left\{\frac{1}{\sin \Pi(b) \sin \Pi(c)} - \frac{1}{\sin \Pi(a)}\right\};$$

in such a way we find:

$$\cos B = \tan \Pi(c) \tan \Pi(a)\left\{\frac{1}{\sin \Pi(a) \sin \Pi(c)} - \frac{1}{\sin \Pi(b)}\right\};$$

$$\cos C = \tan \Pi(a) \tan \Pi(b)\left\{\frac{1}{\sin \Pi(a) \sin \Pi(b)} - \frac{1}{\sin \Pi(c)}\right\}.$$

From these expressions for $\sin A, \cos A, \cos B$, we derive:

$$\sin(A + B) = \sin A \cos B + \cos A \sin B =$$

$$= \tan \Pi(b) \tan^2 \Pi(c) \tan \Pi(a) P \left\{ \frac{1}{\sin \Pi(c) \sin \Pi(a)} - \frac{1}{\sin \Pi(b)} \right\} +$$

$$+ \tan^2 \Pi(c) \tan \Pi(a) \tan \Pi(b) P \left\{ \frac{1}{\sin \Pi(b) \sin \Pi(c)} - \frac{1}{\sin \Pi(a)} \right\} =$$

$$= \tan \Pi(a) \tan \Pi(b) \tan^2 \Pi(c) P \left\{ \frac{1}{\sin \Pi(a)} + \frac{1}{\sin \Pi(b)} \right\} \left\{ \frac{1}{\sin \Pi(c)} - 1 \right\}$$

and at the end

$$\sin(A + B) = \frac{\tan \Pi(a) \tan \Pi(b) P}{\left\{ \frac{1}{\sin \Pi(c)} + 1 \right\}} \left\{ \frac{1}{\sin \Pi(a)} + \frac{1}{\sin \Pi(b)} \right\}.$$

the last from the equation (19) gives:

$$\cos A + \cos(B + C) = \sin B \sin C \left\{ \frac{1}{\sin \Pi(a)} - 1 \right\};$$

inserting here instead of $\sin B, \sin C$ their values from equation (28), and we get:

$$\cos(B + C) = -\cos A + \tan \Pi(c) \tan^2 \Pi(a) \tan \Pi(b) P^2 \left\{ \frac{1}{\sin \Pi(a)} - 1 \right\}$$

or, which is the same

$$\cos(B + C) = -\cos A + \frac{\tan \Pi(b) \tan \Pi(c) P^2}{\frac{1}{\sin \Pi(a)} + 1}.$$

With the help of the previous equations we find:

$$\cos(A + B + C) = \cos A \cos(B + C) - \sin A \sin(B + C) =$$

$$= -\cos^2 A + \frac{\tan^2 \Pi(b) \tan^2 \Pi(c) P^2}{\left\{ \frac{1}{\sin \Pi(a)} + 1 \right\}} \left\{ \frac{1}{\sin \Pi(b) \sin \Pi(c)} - \frac{1}{\sin \Pi(a)} \right\} -$$

$$- \frac{\tan^2 \Pi(b) \tan^2 \Pi(c) P^2}{\left\{ \frac{1}{\sin \Pi(a)} + 1 \right\}} \left\{ \frac{1}{\sin \Pi(b)} + \frac{1}{\sin \Pi(c)} \right\}.$$

$$2\cos^2\frac{(A+B+C)}{2} = \sin^2 A + \frac{\tan^2\Pi(b)\tan^2\Pi(c)P^2}{\left\{\frac{1}{\sin\Pi(a)}+1\right\}} \times$$

$$\times\left\{\frac{1}{\sin\Pi(b)\sin\Pi(c)} - \frac{1}{\sin\Pi(a)}\right\} -$$

$$-\frac{\tan^2\Pi(b)\tan^2\Pi(c)}{\left\{\frac{1}{\sin\Pi(a)}+1\right\}}\left\{\frac{1}{\sin\Pi(b)} + \frac{1}{\sin\Pi(c)}\right\} =$$

$$= \tan^2\Pi(b)\tan^2\Pi(c)P^2 + \frac{\tan^2\Pi(b)\tan^2\Pi(c)}{\left\{\frac{1}{\sin\Pi(a)}+1\right\}} \times$$

$$\times\left\{\frac{1}{\sin\Pi(b)\sin\Pi(c)} - \frac{1}{\sin\Pi(a)} - \frac{1}{\sin\Pi(b)} - \frac{1}{\sin\Pi(c)}\right\} =$$

$$= \frac{\tan^2\Pi(b)\tan^2\Pi(c)P^2}{\left\{\frac{1}{\sin\Pi(a)}+1\right\}}\left\{1 + \frac{1}{\sin\Pi(b)\sin\Pi(c)} - \frac{1}{\sin\Pi(b)} - \frac{1}{\sin\Pi(c)}\right\} =$$

$$= \tan^2\Pi(a)\tan^2\Pi(b)\tan^2\Pi(c)P^2 \times$$

$$\times\left\{\frac{1}{\sin\Pi(a)} - 1\right\}\left\{\frac{1}{\sin\Pi(b)} - 1\right\}\left\{\frac{1}{\sin\Pi(c)} - 1\right\};$$

but it was already proven that, the area of the triangle

$$\triangle = \pi - A - B - C,$$

hence

$$\sin\frac{\triangle}{2} = \frac{1}{\sqrt{2}}\tan\Pi(a)\tan\Pi(b)\tan\Pi(c)P\times$$

$$\times\sqrt{\left(\frac{1}{\sin\Pi(a)}-1\right)\left(\frac{1}{\sin\Pi(b)}-1\right)\left(\frac{1}{\sin\Pi(c)}-1\right)}; \qquad [\text{A}]$$

if a, b, c are so small, that we can use the approximation

$$\frac{1}{\sin\Pi(a)} = 1 + \frac{1}{2}a^2, \qquad \frac{1}{\sin\Pi(b)} = 1 + \frac{1}{2}b^2, \qquad \frac{1}{\sin\Pi(c)} = 1 + \frac{1}{2}c^2,$$

$$\tan\Pi(a) = \frac{1}{a}\left(1 - \frac{1}{6}a^2\right), \qquad \tan\Pi(b) = \frac{1}{b}\left(1 - \frac{1}{6}b^2\right),$$

$$\tan\Pi(c) = \frac{1}{c}\left(1 - \frac{1}{6}c^2\right).$$

then we get:

$$\sin\frac{\triangle}{2} = \frac{1}{2}\sqrt{a^2b^2 - \left(\frac{c^2 - a^2 - b^2}{2}\right)^2} \qquad [\text{B}]$$

or, neglecting the powers of Δ higher then the first we have:

$$\Delta = \sqrt{a^2 b^2 - \left(\frac{c^2 - a^2 - b^2}{2}\right)^2}. \qquad [C]$$

12 CALCULATING THE AREA OF A
CURVED SURFACE

23. Let us define the position of a point in the space of the three rectangular coordinates (Fig. 42), z perpendicular to the planes xy, and y is a perpendicular, drawn from the end z to the axis x, and x is part of the axis x between the origin of the coordinates and the end of y.

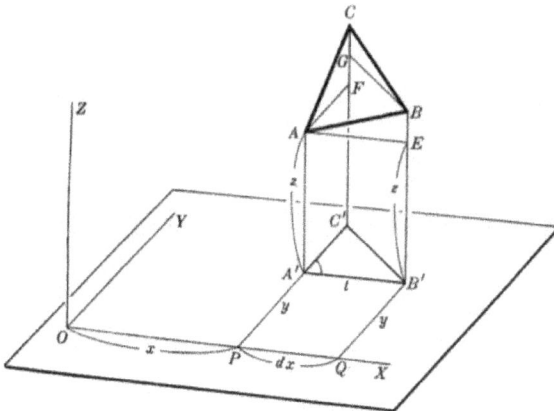

Fig. 42

In order to define an element on the curved space, we take three points $[A, B, C]$ and let the coordinates of the first $[A]$ be x, y, z, the coordinates of a second $[B] - x + dx, y, z + \left(\frac{dz}{dx}\right) dx$, and the coordinates of a third point $[C] - x, y + dy, z + \left(\frac{dz}{dy}\right) dy$. Call t the distance between the top of the two perpendiculars to the axe X, which both are equal to y, and between which part dx of axis x is situated; assuming dx, dy are infinitesimal small we get, based on the equation $(27')$

$$t = \frac{dx}{\sin \Pi(y)}.$$

The distance between the first two points on the curved surface forms triangle with straight lines $[AE]$ and $[EB]$, whose length is equal.

183

$$\frac{dx}{\sin \Pi(y) \sin \Pi(z)}, \quad \left(\frac{dz}{dx}\right) dx.$$

We can consider this triangle as right when its sides are small and the hypothenuse will be the distance between the two first points $[AB]$ on the surface. Hence the square of this distance well be:

$$dx^2 \left\{ \frac{1}{\sin^2 \Pi(y) \sin^2 \Pi(z)} + \left(\frac{dz}{dx}\right)^2 \right\}. \tag{c^2}$$

In such a way we find the square of the distance between the first point and the third $[AC]$

$$dy^2 \left\{ \frac{1}{\sin^2 \Pi(z)} + \left(\frac{dz}{dy}\right)^2 \right\}; \tag{b^2}$$

and the square of the distance of the second point from the third $[BC]$

$$\frac{dx^2}{\sin^2 \Pi(y) \sin^2 \Pi(z)} + \frac{dy^2}{\sin^2 \Pi(z)} + \left\{ \left(\frac{dz}{dy}\right) dy - \left(\frac{dz}{dx}\right) dx \right\}^2. \tag{a^2}$$

The area of a triangle whose sides are the distances from the first to the second point on the curved surface, from the second to the third and from the third to the first, where the sum of the angles is equal to π without a significant error due to the smallness of the sides, will be equal, as a result of the proven above and the values found for the squares of its sides

$$\frac{d^2 S}{dx dy} = \frac{1}{2 \sin \Pi(z)} \sqrt{\left(\frac{dz}{dx}\right)^2 + \frac{1}{\sin^2 \Pi(y)} \left(\frac{dz}{dy}\right)^2 + \frac{1}{\sin^2 \Pi(y) \sin^2 \Pi(z)}}; \tag{29}$$

such an expression for the element of the surface, whose equation is

$$z = f(x,y).$$

We apply this expression to *the sphere*, with a radius r; if the origin of the coordinates is the centre of the sphere, then the equation of the sphere gives

$$\left(\frac{dz}{dx}\right) = -\frac{\cos \Pi(x)}{\cos \Pi(z)}, \quad \left(\frac{dz}{dy}\right) = -\frac{\cos \Pi(y)}{\cos \Pi(z)},$$

after

$$\frac{\cos \Pi(r)}{\sin^2(r)} \frac{\sin \Pi(y) \sin^2 \Pi(x)}{\sqrt{\sin^2 \Pi(x) \sin^2(y) - \sin^2 \Pi(r)}} = \frac{d^2 S}{d\Pi(x) \Pi(y)}.$$

Multiplying by $d\Pi(y)$ and integrating from $\sin \Pi(y) = \frac{\sin \Pi(r)}{\sin \Pi(x)}$, to $\Pi(y) = \frac{\pi}{2}$ we get:

$$\frac{dS}{d\Pi(x)} = \frac{2\pi \sin \Pi(x) \cos \Pi(r)}{\sin^2 \Pi(r)}.$$

Multiplying again by $d\Pi(x)$ and integrating from $\Pi 9x) = \frac{\pi}{2}$, we get:

$$S = \frac{2\pi \cos \Pi(r) \cos \Pi(x)}{\sin^2 \Pi(r)}.$$

This is a surface of a segment of a sphere, between two planes perpendicular to one radius, from which one is passing through the center of the sphere, and the other at a distance x of the center of the sphere. In order to find the surface of the entire sphere, we must to put $x = r$ in the last expression and also to double the value of the expression; in such a way we find the quantity of the surface of the entire sphere is $4\pi \cot^2 \Pi(r)$ or $\pi(e^r - e^{-r})^2$; if r is as small, that we can neglect the higher powers of r, then this expression becomes $4\pi r^2$, as in the ordinary geometry.

24. Suppose

$$\cos \psi = \tan \Pi(r) \cot \Pi(y),$$

$$\cos \Pi(x) = \cos \Pi(r) \sin \psi \sin \varphi$$

and introduce new variables ψ, φ instead of x, y in the expression for the element of the surface of the sphere with the radius r, which we discuss; we find

$$\frac{d^2 S}{d\varphi d\psi} = -\frac{\cos^2 \Pi(r)}{\sin \Pi(r)} \frac{\sin \psi \sqrt{1 - \cos^2 \Pi(r) \sin^2 \psi \sin^2 \varphi}}{1 - \cos^2 \Pi(r) \sin^2 \psi}. \qquad (29a)$$

Multiplying this equation by $8 d\psi d\varphi$ and integrate from $\psi = 0$ to $\psi = \frac{\pi}{2}$ and from $\varphi = 0$ till $\varphi = \frac{\pi}{2}$, in this way we find the surface of the entire sphere. Equalizing the expression, which we find for the surface of the entire sphere with the expression, found above for the same surface, we conclude that

$$\frac{\pi}{2 \sin \Pi(r)} = \int_0^{\frac{\pi}{2}} d\psi \int_0^{\frac{\pi}{2}} d\varphi \frac{\sin \psi \sqrt{1 - \cos^2 \Pi(r) \sin^2 \psi \sin^2 \varphi}}{1 - \cos^2 \Pi(r) \sin^2 \psi} \qquad (30)$$

If we denote $E(\alpha)$ the elliptical integral we get

$$E(\alpha) = \int^{\frac{\pi}{2}0} d\varphi \sqrt{1 - \alpha^2 \sin^2 \varphi},$$

where α is constant under the sign of the integral, then we find with the help of the integral, which represents a sphere:

$$\frac{\pi \alpha}{2\sqrt{1 - \alpha^2}} = \int_0^{\alpha} \frac{x \, dx \, E(x)}{(1 - x^2)\sqrt{\alpha^2 - x^2}};$$

inserting in the integral (30) $\frac{\pi}{2} - R$ instead $\Pi(r)$, we get:

$$\frac{\pi}{2} R = \int_0^{\frac{\pi}{2}} \int_0^{\frac{\pi}{2}} \frac{d\psi d\varphi \sin \psi \sin R}{\sqrt{1 - \sin^2 \psi \sin^2 \varphi \sin^2 R}}.$$

Integrating with respect to ψ in the indicated limits, we find:

$$\pi R = \int_0^{\frac{\pi}{2}} \frac{d\varphi}{\sin\varphi} \log\left(\frac{1+\sin\varphi\sin R}{1-\sin\varphi\sin R}\right),$$

when we put $\Pi(x)$ instead φ we get

$$\pi R = \int_0^\infty dx \log\left\{\frac{e^{2x}+1+2e^x\sin R}{e^{2x}+1-2e^x\sin R}\right\};$$

by integration by parts we get the following equation:

$$\frac{1}{4}\pi\frac{R}{\sin R} = \int_0^\infty \frac{(e^{2x}-1)e^x x\,dx}{e^{4x}+2e^{2x}\cos 2R+1}. \tag{31}$$

for $R = \frac{\pi}{2}$ this equation gives:

$$\frac{1}{8}\pi^2 = \int_0^\infty \frac{x\,dx\,e^x}{e^{2x}-1}.$$

It is easy to prove equation (31) for $\cos 2R > 1$.
We in fact have:

$$\int_0^\pi d\psi \log\cot\psi = 0,$$

from where follows, that for any number a we have

$$\int_0^\pi d\psi \log\left(e^a \cot\frac{1}{2}\psi\right) = a\pi.$$

Applying this equation, putting in it $e^a \cot\frac{1}{2}\psi = e^x$;

$$\int_{-\infty}^{+\infty} \frac{x\,dx}{e^{x-a}+e^{-x+a}} = \frac{1}{2}\pi a;$$

this equation can be easily rewritten in the following way:

$$\int_0^\infty \frac{(e^x-e^{-x})x\,dx}{e^{2x}+(e^{2a}+e^{-2a})+e^{-2x}} = \frac{1}{2}\frac{\pi a}{e^a-e^{-a}},$$

from where we find again the equation (31), inserting $a\sqrt{-1}$ instead of a.

13 CALCULATING THE VOLUME OF BODIES

25. If for coordinates we take the arcs of a limiting circle, one ζ drawn from a given point in the plane, intersecting the plane xy in a straight line parallel to the axis x and to which arc ζ is parallel, the extension on the side of the parallelism is an axis, the other η in the plane x, y drawn from the end of ζ to axis x, which also serves with the axis to that arc. If at the end for the third coordinate we accept ξ, a segment of the axis x from the origin of the coordinate to the end of η, then the element of the volume P should be $d\xi d\eta d\zeta$. Hence

$$d^3 P = d\xi d\eta d\zeta.$$

Suppose that $\zeta = \cot \Pi(z)$, where z is a perpendicular from a given point on the plane xy, we get:

$$\left(\frac{d^3 P}{d\xi d\eta dz} \right) = \frac{1}{\sin \Pi(z)}.$$

From the equation of the limiting circle we derive

$$e^{-p} = \sin \Pi(z),$$

where p is the distance of intersection of the arc ζ with the plane xy to the end of perpendicular z.

Noticing that as a result of the equation of the limiting circle and the expression of the arcs of the limiting circle depending on the ordinates:

$$\cot \Pi(y) = \eta e^{-p},$$

$$e^{\xi - x} = \sin \Pi(z) \sin \Pi(y),$$

we find:

$$\frac{d\eta}{dy} = \frac{1}{\sin \Pi(y) \sin \Pi(z)}; \quad dx = d\xi;$$

from where it follows, that

$$\left(\frac{d^3 P}{dx dy dz} \right) = \frac{1}{\sin \Pi(y) \sin^2 \Pi(z)}. \tag{31a}$$

187

Multiplying this expression by dx and integrating it from $x = 0$, we get:

$$\left(\frac{d^2 P}{dydz}\right) = \frac{x}{\sin \Pi(y) \sin^2 \Pi(z)}.$$

Multiplying the same expression by dy and integrating it from $y = 0$, we get:

$$\left(\frac{d^2 P}{dxdz}\right) = \frac{\cot \Pi(y)}{\sin^2 \Pi(z)}.$$

Multiplying at the end by dz and integrating it from $z = 0$, we get:

$$\left(\frac{d^2 P}{dxdy}\right) = \frac{1}{8 \sin \Pi(y)} \{e^{2z} - e^{-2z} + 4z\}.$$

If multiplying penultimate of these expression by $dxdz$ and after integrate from the beginning regarding z from $z = 0$ till the values of z taken from the equation $\sin \Pi(r) = \sin \Pi(x) \sin \Pi(z)$, and after regarding x from $x = 0$ till $x = r$, and if multiplying the result by 8, in order to get the value of the entire sphere, then we find the volume of the whole sphere $= \frac{1}{2}\pi\{e^{2r} - e^{-2r} - 4r\}$, for r extremely small gives $\frac{4}{3}\pi r^3$ as in the ordinary geometry.

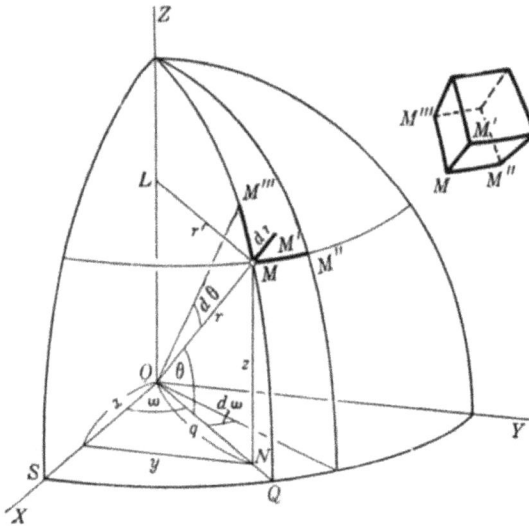

Fig. 43

26. In order to express the element of a volume in polar coordinates, we call r a distance from the origin of the coordinates to the point $[M]$ in space, rectilinear coordinates x, y, z (Fig. 43).

Let call the straight line q $[ON]$, drawn from the origin of the coordinates till the end z, θ is the angle between r and q, ω is the angle between q and the axis of positive x. Suppose that $\Pi(x) = X$, $\Pi(y) = Y$, $\Pi(z) = Z$, $\Pi(r) = R$, $\Pi(q) = Q$. We draw through the given point of the plane perpendicular to axis z. Let r' is the straight line $[ML]$ drawn in that plane from a given point to the axis z and suppose that $\Pi(r') = R'$.

Drawing around the origin of the coordinates as center of the sphere radius r. The plane xy intersects the sphere in a big circle, whose circumference will be according to the proven above will be:

$$2\pi \cot R.$$

Part of this circle $[QS]$ which is between two planes, drawn through axis z and inclined one to the other under the angle ω should be

$$\omega \cot R.$$

Circumference of the circle resulting from the intersection the same sphere passing through the given point and perpendicular to axis z, will be:

$$2\pi \cot R'.$$

And a part of this circumference which lies between two planes, drawn through the axis z and inclined under the angle ω, should be:

$$\omega \cot R'.$$

The change of $[MM'']$, done in the last arc, while increasing the angle ω by $d\omega$, should be

$$d\omega \cot R'.$$

The right triangle $[MOL]$, where hypotenuse is r, one of the sides of the right angle r' and the opposite angle r' is equal to $\frac{\pi}{2} - \theta$, gives (equation 12) $\tan R' \cos\theta = \tan R$, from where it follows that

$$d\omega \cot \mathbf{R'} = d\omega \cot \mathbf{R} \cos\theta.$$

The circumference of the circle due to intersection of the same sphere and the plane, going through axis z, is equal to

$$2\pi \cot \mathbf{R},$$

and the arc of this circle, which correspond to angle θ at the center will be:

$$\theta \cot R;$$

from where it follows, that the change of the arc $[MM'']$, which corresponds to the augmentation $d\theta$ to the angle θ should be:

$$d\theta \cot R.$$

If all augmentations are infinitesimal then the element of the volume is expressed, as in the ordinary geometry, by the product of three lines, perpendicular among themselves

$$dr, d\omega \cos\theta \cot \mathbf{R}, d\theta \cot \mathbf{R},$$

because we can consider this volume as a prism.

Hence, the element of the volume is expressed in polar coordinates:

$$dr\, d\omega\, d\theta \cos\theta \cot^2 R = d^3 P$$

or replacing $\cot^2 \mathbf{R}$ with its value in r:

$$d^3 P = \frac{1}{4} dr\, d\omega\, d\theta \cos\theta (e^r - e^{-r})^2.$$

Integrating firstly with respect to r from $r = 0$, we get:

$$d^2 P = \frac{1}{8} d\omega\, d\theta (e^{2r} - e^{-2r} - 4r).$$

For the sphere which center is in the origin of the coordinates, r does not depend on θ and ω.

Integrating with respect to ω from $\omega = 0$ till $\omega = 2\pi$ and with respect to θ from $= 0$ till $\theta = \frac{\pi}{2}$ and multiplying the result by 2, we get volume of the whole sphere: $\frac{1}{2}\pi(e^{2r} - e^{-2r} - 4r)$, as above.

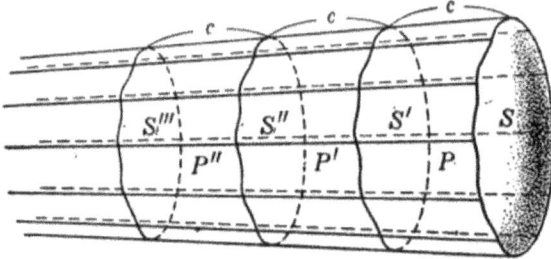

Fig. 44

27. Consider now (Fig. 44) part S of the surface of the limiting sphere, bounded by a closed line, we draw from different points of that borderline straight lines to the axis of the sphere; such straight lines form a surface, which we call by resemblance conical and which is extended infinitely in both sides, but on which we will study only part of the limiting sphere, infinitely extended on the side of parallelism of the axes of the limiting sphere. Let S' be part of another limiting sphere with axes, parallel to the previous and facing to the same side; the part which is inside of a conical surface. S, S' and part between the two limiting spheres enclose a volume, finite in all sides, which we suggest to determine. We call c the part of the axis between two limiting spheres. After that we repeat, replicate the straight line c several times on the one of the axes of the limiting sphere, passing through one of the point on the curve, limiting S, starting with the point, where this axis intersects S'. Draw through points of division of the limiting spheres with axes, parallel to the axes of the first two and facing the same side. Let S'', S''' and so on are parts of these limiting spheres inside the conical surfaces.

On the base of what we proved here about the arcs of limiting circle, with similar arrangement, and also about of parts of limiting spheres, follows that

$$S' = Se^{-2c}, S'' = Se^{-4c}, S''' = Se^{-6c}.$$

We call P, P', P'' and so on the volumes inside the conical surface between S, S', between S' and S'' and so on and notice that volumes P, P', P'' and so on should be proportional to the surfaces S, S', S'' and so on. So we should have $P = CS$, where C depends only on c, from where follows, that:

$$P' = CS' = CSe^{-2c}, \quad P'' = CS'' = CSe^{-4c}.$$

The sum $\sum_{n=0}^{\infty} P^{(n)}$ will be a volume inside the conical surface, whose base will be S and which extends unlimitedly to the side of the parallelism of the straight derivatives. Let that volume be K; we get:

$$K = \frac{CS}{1 - e^{-2c}}.$$

This quantity should not depend on c and for it we require that:

$$C = (1 - e^{-2c})A,$$

where A is an abstract number and since the volume unit is arbitrary, then we accept that $C = \frac{1}{2}(1 - e^{-2c})$ so that the volume P be expressed in the formula

$$P = \frac{1}{2} S(1 - e^{-2c}),$$

turns into $P = cS$ for infinitesimal c expression, which according to that, how it is expressed in the ordinary geometry the volume of prism with base S and hight c.

It is possible as well to accept for the volume element a volume inside the canonical surface, formed by the axes of the limiting sphere, drawn through all points of the curve, limiting a part of the given surface which is infinitesimally small in all directions.

The big number of different expressions for the element of the same geometrical quantity gives us means for comparing the integrals, means, which are useful in the theory of the definite integrals.

14 CONCLUSION

28. Having shown so far how to calculate the length of curved lines, magnitude of surfaces and the magnitude of the volume of bodies, we are right to say, that pangeometry is a complete theory of geometry. Glancing at equations, which show dependence of the angles and sides of rectilinear triangles, it is enough to show, that starting with these equations of pangeometry we do analytical calculations, which replace and generalize the analytical method of ordinary geometry. We can now start the exposition of pangeometry with the equations (19) and even to try to replace these equations with others, which would have been expressed the dependence of the sides and angles of any rectilinear triangle. But in the last case it should be proved that these new equations are consistent with the basic geometrical notions. Equations (19) being introduced with the help of these geometrical notions, should be in accordance with them.

So, equations (19) serve as the basis of geometry in some general kind because they do not depend on suggestions, that the sum of the three angles in each rectilinear triangle is equal to two right angles.

Pangeometria, as it is explained here, is based on the foundations as we saw provides means for calculations of different geometrical quantities and proves also, that the accepted in the ordinary geometry explicitly or implicitly, that the sum of the three angles each rectilinear triangle is constant, is not a necessary consequence of our notions of space. Only experiment can confirm the trustfulness of this assumption, for example the measurement of the three angles of rectilinear triangle, this measurement which can be done in different ways. It is possible to measure the three angles in a triangle, build on an artificial plane, or three angles of one triangle in space. In the latter case it is preferable to consider triangles, which sides are very big, because according to the theory of pangeometry, the difference of the sum of three angles of the triangle with two right angles is as greater as the sides are bigger.

Let (Fig. 45) r be a radius of a circle, A is an angle at the center between the radii opposite to the chords and is equal to r. Let call p a perpendicular, drawn from the center of the circle to the chord r, which it divides into two equal parts. Consider one from the rectangular triangles, whose perpendicular sides are p and $\frac{r}{2}$, and the hypotenuse is r.

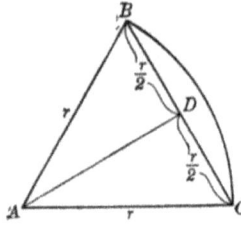

Fig. 45

According to the general equation (13), in this triangle we will have:

$$\sin\frac{A}{2}\tan\Pi\!\left(\frac{r}{2}\right)=\tan\Pi(r),$$

equation, which together with the identical equation gives:

$$\tan\Pi(r)=\frac{\sin^{2}\Pi\!\left(\frac{r}{2}\right)}{2\cos\Pi\!\left(\frac{r}{2}\right)}$$

gives:

$$\sin\frac{A}{2}=\frac{1}{2}\sin\Pi\!\left(\frac{r}{2}\right).$$

In the ordinary geometry we have:

$$A=\frac{\pi}{3}.$$

Let the real measurement gives:

$$A=\frac{2\pi}{6+K},$$

where K is a positive number. Hence we should have

$$\sin\!\left(\frac{\pi}{6+K}\right)=\frac{1}{2}\sin\Pi\!\left(\frac{r}{2}\right).$$

If K and r are given, then it is possible to derive from this equation the following value $\Pi(\frac{r}{2})$; after that we find the angle of parallelism $\Pi(x)$ for each straight line x.

The distance between the celestial bodies gives us possibility to observe the angles of triangles, whose sides are very large.

Let call α geocentric latitude of a motionless (distant) star (Fig. 46.) in a given epoch, and β is another geocentric latitude of the same star, which is responsible for time, when the Earth is again in the plane perpendicular to the ecliptic and drawn through the first location of the star. Let $2a$ be a distance between these two positions of the Earth, δ is angle, under which

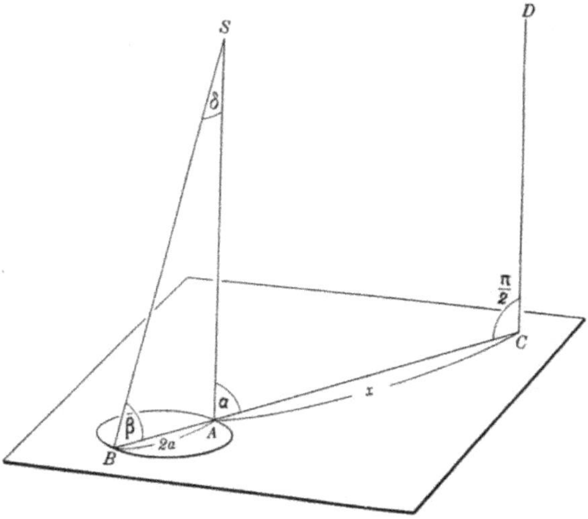

Fig. 46

we see the distance $2a$ from the star. If the angles α, β, δ do not satisfy the equation $\alpha = \beta + \delta$, then this will be a sign that the sum of the three angles of this triangle is not equal to two right angles. We can choose a star, that $\delta = 0$ and we can always suggest that exists line x for which

$$\Pi(x) = \alpha.$$

If $\delta = 0$, then the straight lines, drawn from two positions of the Earth to the star can be regarded as parallel. As a result should be $\beta = \Pi(x+2a)$, from where it follows in accordance with proven above equation, that

$$\tan \frac{\alpha}{2} = e^{-x}, \quad \tan \frac{\beta}{2} = e^{-x-2a}.$$

every time, when α and β in the observations of the stars, for which $\delta = 0$ will be different, the last two equations will give x and a, expressed by a line, taken for the unit in pangeometry. In such a way, knowing the angle of parallelism $\Pi(x)$ for a given line x, we can calculate the angle $\Pi(y)$ for every line y.